Technology and Theology: How AI is Impacting Religion

Andrew Bloom

Published by Ethics & Innovation Press, 2024.

While every precaution has been taken in the preparation of this book, the publisher assumes no responsibility for errors or omissions, or for damages resulting from the use of the information contained herein.

TECHNOLOGY AND THEOLOGY: HOW AI IS IMPACTING RELIGION

First edition. November 25, 2024.

Copyright © 2024 Andrew Bloom.

ISBN: 979-8991958899

Written by Andrew Bloom.

Dedication

To Michal — From the day we met on the quad in college, I knew I had been blessed. You have been a blessing every day since. Your love, support, and partnership have been my foundation. Your presence in every step of this journey has made all the difference, and I couldn't have done it without you. I love you to infinity and beyond.

To Daniel, Maya, and Lia — this book is a reflection of your unique interests, the challenges you've faced, and the many conversations we've shared. Being your father is the greatest gift of all, and you inspire me every day.

To My Parents - Erwin and Gail Bloom - Thank you for your love and support and for igniting the spark of Judaism within me.

To My Sister - Lauren Bloom - your work within the Jewish field is an inspiration.

To Rabbi Sidney Zimelman — your wisdom and our chavrutah studies have left an indelible mark on my learning and growth. You are a true friend and Rabbi's Rabbi.

To Ellis Slater (GT) — my mentor, guide, and friend for over 25 years — Slàinte! Your insight, guidance, and friendship have illuminated my path on this journey of growth. In the spirit of JON: "All shall be well, and all shall be well, and all manner of things shall be well.

Introduction

As a Conservative Rabbi and an AI ethicist, I find myself at the crossroads of two powerful forces that shape our world: the ancient wisdom of religious traditions and their interpretations alongside the transformative, ever-evolving technological innovations. At first glance, theology and technology might seem worlds apart, perhaps even at odds. Theology, with its millennia of reflection on the human condition, moral responsibility, and our relationship with God seemingly could appear hesitant to embrace the rapid pace of technological change. This change evolved even faster since the COVID pandemic. Conversely, technology, particularly artificial intelligence (AI), often seems focused on the future, progress, and efficiency, moving forward without looking backward unless it needs historical data for advanced learning.

Yet, from my vantage point, these two realms are not as divergent as they might seem. The Torah, while grounded in ancient truths, and beliefs has always remained relevant by adapting to the new realities and knowledge that each generation lives within. The teachings of our sages, from the early Talmudic scholars to modern-day commentators, constantly grapple with the ethical and moral dilemmas of their and our times. Similarly, AI represents more than a new set of tools; it is a profound reshaping of how we interact with each other, how we solve problems, and how we perceive the world around us. The influence of AI extends beyond the mechanical or computational—it touches on questions of free will, morality, and what it means to be human. In this way, AI invites us into a dialogue that theology is well-equipped to both participate in and guide.

In this book, I intend to explore the intersection of AI and theology through the lens of Jewish thought, drawing from our biblical and rabbinic teachings to address modern technological challenges. My journey into this field which began in 1996 at the Schechter Institute of Rabbinical Studies and the Online and Internet Department of the Jerusalem Post has been deeply influenced by my ever evolving theological and technological interests, including research into the ethics of AI. These identities are often seen as separate, yet they inform and enrich each other in profound ways. The ethical dilemmas I encounter in AI echo the moral questions raised by our Torah and rabbinic literature. As a rabbi, I am compelled to draw from the rich tapestry of Jewish tradition to provide ethical guidance in navigating the complexities of AI.

To begin with, we must ask: why should religious communities concern themselves with AI? Why engage with technology that is often seen as secular, if not a challenge to the authority of religious wisdom? The answer, I believe, lies in the Jewish tradition itself. In the Talmud, there is a famous teaching: "Turn it, and turn it again, for everything is in it." [1] This refers to the Torah and its capacity to provide insight into all aspects of life, from the mundane to the transcendent. If we believe that everything is indeed within the Torah, then surely, technology—like agriculture, medicine, and law before it—falls under the purview of our religious inquiry. These sorts of real-life issues when the COVID pandemic challenged us to use electronic devices like ZOOM on Shabbat and Chagim/Holidays.

Consequently, technology, and AI in particular, has become a central force in shaping our lives, and it does so in ways that raise deep theological questions. Consider, for example, the concept of free will, a core principle of Jewish theology. We are taught that human beings are endowed with the ability to make choices, and these choices are what give our lives moral meaning. But what happens when we delegate decision-making to machines? AI algorithms now guide critical decisions, from healthcare to criminal justice to financial markets. How do we ensure that these machines reflect the values of justice and righteousness that are central to our tradition? Are we at risk of creating a system where moral accountability is diffused, or worse, undermined?

Another profound theological challenge arises when we consider the nature of human uniqueness. The Torah teaches that human beings are created "b'tzelem Elohim"[2]—in the image of God. This divine spark sets us apart from the rest of creation. But what happens when machines begin to exhibit traits that we once considered uniquely human? AI systems now have the capacity to learn, adapt, and even engage in creative processes. While these systems are far from sentient, they push the boundaries of what we once thought only humans could do. Does this erode our sense of human exceptionalism? Or does it challenge us to rethink what it means to bear the image of God in a world where technology increasingly mirrors human intelligence?

These are not just abstract questions. They have real-world implications for how we structure society and live our lives. AI is already shaping our economic systems, our political institutions, and our interpersonal relationships. As a rabbi, I believe it is imperative that we bring our theological wisdom to bear on these changes. Jewish tradition, with its deep commitment to justice, compassion, and communal responsibility, offers a critical ethical framework for navigating the promises and complexities of AI.

Consider the biblical story of the Tower of Babel.[3] In this narrative, human beings come together to build a great tower, seeking to reach the heavens and make a name for themselves. But their efforts are thwarted by God, who confounds their language and scatters them across the earth. This story has often been interpreted as a cautionary tale about the dangers of technological hubris, the belief that human ingenuity can transcend divine boundaries. Yet, there is another layer to this story that is relevant to our discussion of AI. The builders of the tower were unified in their purpose, but they lacked a moral foundation. Their ambition was not tempered by ethical considerations, and this ultimately led to their downfall.

In our contemporary world, AI represents a new kind of tower—an awe-inspiring technological achievement with the potential to reshape human civilization. But like the builders of Babel, we face the risk of moving forward without sufficient ethical grounding. We must ask ourselves: What kind of society do we want AI to help us build? Will it be a society that reflects the values of justice, kindness, and humility, or one that is driven solely by efficiency and power?

As we embark on this journey of integrating AI into our lives, we must remember the wisdom of our tradition: technology is not inherently good or evil, but it is a tool that reflects the values of those who wield it. In Jewish thought, the concept of "Tikkun Olam"[4], repairing the world, calls us to use our creative capacities for the betterment of society. AI has the potential to be a powerful force for Tikkun Olam, addressing challenges like poverty, disease, and climate change. But it also carries the potential for harm, particularly if it exacerbates inequality, dehumanizes individuals, or undermines our sense of moral responsibility and adherence to God's statutes and commandments.

The chapters that follow will delve deeper into these questions, examining specific areas where AI and theology intersect. We will explore how AI challenges our understanding of free will, how it reshapes our concepts of justice and fairness, and how it impacts our sense of communal responsibility. We will also look at the opportunities AI presents for enhancing religious life, from improving access to education to creating new forms of spiritual engagement.

Throughout this book, my goal is not to provide definitive answers but to open a conversation—a dialogue between the ancient wisdom of our biblical and rabbinic traditions and the new realities of the AI era. I believe that theology has much to offer this conversation, providing a moral compass for navigating the ethical complexities of AI. At the same time, I believe that engaging with AI can deepen our theological understanding, pushing us to reconsider long-held assumptions and uncover new insights into the nature of humanity, divinity, and the world we inhabit.

In the words of the prophet Micah, "What does the Lord require of you? To act justly, to love mercy, and to walk humbly with your God."[5] As we stand on the threshold of an AI-driven future, these words serve as a reminder of our ethical responsibilities. AI offers incredible possibilities, but it is up to us to ensure that it is guided by justice, mercy, and humility. In doing so, we can fulfill our role as co-creators in the ongoing process of Tikkun Olam, using technology not to replace our humanity, but to elevate it.

Footnotes

[1] Pirkei Avot 5:22 – This Mishnah teaches the importance of continuous study and engagement with the Torah, emphasizing that every aspect of life is encompassed within its teachings.

[2] Genesis 1:27 – "So God created mankind in his own image, in the image of God he created them; male and female he created them." This verse underscores the inherent dignity and uniqueness of human beings.

[3] Genesis 11:1-9 – The narrative of the Tower of Babel, where humanity's attempt to build a tower reaching the heavens leads to the confusion of languages and the scattering of people across the earth.

[4] Tikkun Olam – A Hebrew phrase meaning "repairing the world," representing the Jewish ethical mandate to improve society and address social injustices.

[5] Micah 6:8 – "He has shown you, O mortal, what is good. And what does the LORD require of you? To act justly and to love mercy and to walk humbly with your God." This verse encapsulates the ethical expectations placed upon individuals.

Chapter 1: What is AI? Understanding the Technology

We will in this chapter be delving into the theological implications of AI. The book provides a historical and conceptual overview and introduces readers to the foundational concepts and evolution of artificial intelligence (AI) from my perspective as a rabbi in the conservative movement and from my perspective as an AI ethicist. The chapter traces AI's journey from ancient philosophical musings to its status as a transformative technology, explaining key methodologies like machine learning and neural networks. It highlights significant milestones, such as Alan Turing's Turing Test and the Dartmouth Conference, while contrasting early Symbolic AI with the rise of machine learning and deep learning. By categorizing AI into Narrow AI, General AI, and Super AI, the chapter illustrates real-world applications across various industries and emphasizes the ethical dimensions intertwined with technological advancements. This exploration sets the stage for deeper discussions on the intersection of faith, morality, and technology in the chapters to follow.

Chapter 2: The Intersection of AI and Faith

Following the introduction, Chapter 2 explores the Intersection of AI and Faith. This chapter introduces the central themes of the book, discussing why it is pertinent to examine AI through a theological lens. From a Jewish perspective, it touches upon how traditional teachings engage with ethics and technology, and the broader quest for understanding the divine in an age of rapid technological advancement. It also emphasizes the universal relevance of these questions, making connections to various religious and secular viewpoints.

Chapter 3: Theological Foundations: Divine Creation and Human Innovation

Chapter 3 delves into the Theological Foundations, particularly focusing on concepts such as *B'tzelem Elohim* (Image of God) in Jewish thought. It examines how these ideas relate to human creativity and technological innovation, drawing parallels with other religious traditions' views on creation. This chapter discusses the theological significance of humans creating intelligent machines and the responsibilities that come with such power, including stewardship and ethical considerations.

Chapter 4: Ethical AI: Moral Imperatives and Religious Teachings

In Chapter 4, the focus shifts to Ethical AI, exploring how key Jewish ethical principles like *Tikkun Olam* (repairing the world) apply to AI development. The chapter broadens the discussion by incorporating ethical guidelines from various religions and secular philosophies, presenting a multifaceted view of AI ethics. It addresses practical implications, offering guidance on how these ethical frameworks can inform AI policies and practices to ensure technology serves the greater good.

Chapter 5: AI and Free Will: Autonomy, Determinism, and Divine Agency

Chapter 5 tackles the complex relationship between AI and Free Will. It delves into theological debates surrounding free will in Judaism and examines how AI challenges traditional notions of human autonomy. By comparing interpretations from different religions, the chapter explores the implications of AI-driven decision-making on human agency and moral responsibility, raising important questions about the balance between technological determinism and divine agency.

Chapter 6: The Soul and Consciousness: Can AI Possess Spiritual Attributes?

In Chapter 6, the book explores the profound question of whether AI can possess spiritual attributes. From a Jewish perspective, it investigates beliefs about the soul (*Neshama*) and consciousness, assessing whether these can be attributed to artificial entities. The chapter includes interfaith comparisons, drawing insights from Christianity, Islam, Buddhism, and other faiths. It engages with philosophical questions about the potential for AI to attain a form of consciousness or spiritual essence, encouraging readers to ponder the deeper implications of intelligent machines.

Chapter 7: AI in Religious Practice: Enhancing and Transforming Worship

Chapter 7 examines how AI is Integrating into Religious Practice. It looks at specific examples within Judaism, such as virtual prayer services which continue to accelerate since the COVID pandemic, and Torah study aids enhanced by AI. The chapter also highlights how other religions are utilizing AI to enhance worship, build communities, and facilitate religious education. It assesses the benefits, such as deepened spiritual experiences, alongside challenges like the risk of depersonalization, providing a balanced view of AI's role in contemporary religious life.

Chapter 8: Social Justice and AI: Addressing Inequality and Promoting Equity

In Chapter 8, the discussion centers on Social Justice and AI. It explores Jewish social teachings, including principles like *Tzedek* (justice) and *Gemilut Hasadim* (acts of loving kindness), in the context of AI-driven societal changes. The chapter expands to include perspectives from various religious and secular frameworks, addressing AI's impact on social inequality and equity. It offers recommendations aimed at ensuring AI contributes to a more just and equitable society, aligning technological advancements with ethical imperatives.

Chapter 9: The Future of Work and Human Purpose: AI's Impact on Vocation and Meaning

Chapter 9 investigates The Future of Work and Human Purpose in an AI-augmented world. Drawing from Jewish insights on *Avodah* (modern Hebrew interpretation) and its significance in human identity, the chapter reflects on how AI might redefine careers, skills, and the search for meaning. It incorporates comparative views from different traditions on work and purpose, analyzing how AI's integration into various industries could influence our understanding of vocation and personal fulfillment.

Chapter 10: AI Governance and Religious Leadership: Shaping Policies with Ethical Wisdom

Chapter 10 addresses AI Governance and Religious Leadership. It highlights the role of rabbis and other religious leaders in guiding ethical AI governance, emphasizing the importance of integrating religious ethical wisdom into policy-making. The chapter examines interfaith collaboration, showcasing how diverse religious voices can contribute to shaping global AI policies. It suggests frameworks for ensuring that governance structures reflect ethical and moral considerations rooted in various faith traditions.

Chapter 11: Hope and Caution: Envisioning a Future Where AI and Faith Coexist Harmoniously

In Chapter 11, the book presents a balanced view through Hope and Caution. It envisions scenarios where AI and religious values synergize to enhance human flourishing, highlighting optimistic outlooks where technology supports spiritual and social well-being. Simultaneously, it discusses potential pitfalls and ethical dilemmas that must be navigated carefully to prevent negative outcomes. This chapter encourages a thoughtful and proactive approach to integrating AI with faith-based values.

Chapter 12: AI, Privacy, and Surveillance: A Jewish Ethical Response

In Chapter 12, the discussion shifts to the growing concerns around privacy and surveillance in the digital age, particularly how AI-driven technologies have expanded the scope and scale of surveillance. From facial recognition to data mining, AI systems collect vast amounts of information about individuals, raising ethical and moral questions. This chapter explores these concerns through a Jewish ethical lens, focusing on the principles of *Shmirat HaLashon* (guarding one's speech) and *Lashon Hara* (harmful speech/gossip), and their broader implications for privacy. Drawing on Jewish teachings about human dignity and the value of *Tzelem Elohim*, this chapter discusses the balance between the benefits of surveillance (such as security) and the protection of individual privacy. The chapter also touches on the responsibilities of companies, governments, and individuals to create ethical frameworks for AI surveillance, ensuring they respect human dignity while preventing harm.

Chapter 13: AI and Education: Transforming Torah Study and Religious Learning

Chapter 13 explores the profound impact AI is having on education, particularly within religious contexts like Torah study and Jewish learning. The chapter highlights how AI tools are being used to enhance the study of sacred texts through features like automated commentary analysis, interactive Torah study applications, and personalized learning pathways. These technological advancements provide new ways for students to engage with religious texts, enabling deeper learning and accessibility. However, the chapter also considers potential drawbacks, such as the risk of over-reliance on AI tools for religious instruction, which may compromise traditional, human-guided mentorship in Jewish education. The chapter integrates Jewish teachings on the value of *Chevruta* (partnership in study) and *Masorah* (the transmission of tradition) to offer a nuanced view of AI's role in religious education, advocating for a balanced approach that blends human wisdom with technological innovation.

Chapter 14: AI and Interfaith Dialogue: Building Bridges Between Traditions

In Chapter 14, the focus broadens to the role AI can play in fostering interfaith dialogue and collaboration. AI technologies can be harnessed to create platforms for shared learning, facilitate conversations, and promote understanding between different religious traditions. This chapter examines how AI-driven systems can analyze vast religious texts across multiple faiths, uncovering common ethical themes and spiritual insights that can serve as a foundation for interfaith dialogue. Jewish teachings on *Kiddush Hashem* (sanctification of God's name) and *Darchei Shalom* (ways of peace) provide a theological basis for using technology to promote peace and mutual understanding. The chapter includes case studies of successful interfaith AI projects, showcasing how Jewish, Christian, Muslim, and other religious leaders have worked together to address global challenges such as poverty, climate change, and social justice through AI-powered initiatives. The chapter concludes by offering guidelines on how religious communities can continue to use AI as a tool for fostering unity and collaboration across faith boundaries, reinforcing the importance of ethics and shared values in this endeavor.

Chapter 15: Thoughts on Bridging Technology and Theology for a Better Tomorrow

The conclusion synthesizes the key insights from each chapter, emphasizing the profound and transformative relationship between AI and religious thought. As explored throughout the book, AI offers both incredible opportunities and significant ethical challenges. Whether in the realms of social justice, privacy, education, governance, or interfaith dialogue, the intersection of technology and theology reveals how ancient wisdom can inform and guide modern technological advancements.

Chapter 1: What is Artificial Intelligence?

Artificial Intelligence (AI) has evolved from a nascent field of theoretical inquiry into one of the most transformative technologies in the 21st century.[1] Defined broadly as the capability of machines to perform tasks that typically require human intelligence,[2] AI encompasses a wide range of techniques, from rule-based algorithms to deep learning neural networks.[3] This article will explore the historical milestones that have defined the development of AI, break down the various categories of AI, and provide specific examples to illustrate these categories.

The Origins of Artificial Intelligence

AI's origins can be traced back to ancient myths and the philosophical musings of figures such as Aristotle and Descartes,[4] who imagined mechanical beings capable of thinking.[5] However, AI as a scientific discipline began in the 20th century, spurred by advancements in mathematics, logic, and computer science.[6]

One of the earliest formal explorations into AI was made by British mathematician Alan Turing, who in 1950 proposed the famous Turing Test in his paper "Computing Machinery and Intelligence."[7] Turing posited that if a machine could engage in a conversation indistinguishable from a human, it could be said to possess intelligence.[8] This idea laid the foundation for the conceptual and practical development of AI.[9]

The Early Days: Symbolic AI and Good Old-Fashioned AI (GOFAI)

AI research officially began in 1956 during the Dartmouth Conference,[10] where John McCarthy, Marvin Minsky, Nathaniel Rochester, and Claude Shannon came together to discuss the potential of machines to exhibit intelligence.[11] The goal was to mimic human reasoning through symbolic logic, a subfield now referred to as Symbolic AI or Good Old-Fashioned AI (GOFAI).[12]

An example of early AI programs such as The Logic Theorist (1955) by Allen Newell and Herbert A. Simon were designed to prove mathematical theorems.[13] Their program demonstrated that machines could reason using logic, solving problems in ways that resembled human thought.[14]

Despite initial optimism, this early period of AI research encountered significant limitations.[15] Symbolic AI struggled with handling ambiguity and the complexities of real-world environments,[16] leading to the first "AI winter," a period of reduced funding and interest in the field.[17] On a personal note, it was C3PO and his understanding of over 7 million forms of communication that got me interested in the subject.

The Rise of Machine Learning: A New Paradigm

The AI landscape shifted in the 1980s with the emergence of Machine Learning (ML), a subset of AI that allows machines to learn patterns from data without being explicitly programmed.[18], Unlike symbolic AI, ML systems rely on statistical techniques to "learn" from large datasets.[19]

Example: One of the earliest successes in machine learning came with decision trees, where programs could classify data points based on learned decision rules.[20] This method evolved into more complex models like Random Forests and Support Vector Machines (SVMs).[21] These models themselves are worth further individual inquiry.

Neural networks, initially inspired by the structure of the human brain,[22] also emerged during this period.[23] However, due to computational limitations, neural networks did not achieve widespread success until decades later.[24]

The Dawn of Deep Learning

The mid-2000s brought the resurgence of neural networks in the form of deep learning,[25] thanks to increased computational power, larger datasets, and algorithmic breakthroughs.²⁶ Deep learning uses multi-layered neural networks to perform complex tasks such as image recognition, natural language processing, and autonomous driving.²⁷

The 2012 breakthrough by 2024 Nobel Prize Winner Geoffrey Hinton and his team, who developed a deep neural network that significantly improved performance in the ImageNet competition,²⁸ marked a turning point in AI.²⁹ This success demonstrated that machines could surpass human abilities in specific domains, such as object recognition.³⁰

The advent of deep learning has since driven rapid advances across many fields.³¹ AI models like GPT-3 (Generative Pre-trained Transformer 3)³² and AlphaGo, a program developed by DeepMind that defeated the world champion Fan Hui in the complex board game Go,³³ have shown the profound capabilities of AI when combined with deep learning.³⁴

Categories of AI: Narrow, General, and Super AI

AI can be broadly divided into three categories, each with distinct characteristics and applications: Narrow AI, General AI, and Super AI.³⁵

1. **Narrow AI (Weak AI)**
   Narrow AI refers to systems designed to perform a specific task or a limited range of tasks.³⁶ These systems do not possess generalized intelligence and cannot perform tasks outside of their predefined functions.³⁷
   The following below are examples of Narrow AI

   - Siri and Alexa: Virtual assistants that use natural language processing to understand and respond to user queries.³⁸
   - Recommendation engines: Used by companies like Netflix and Amazon to suggest content based on user preferences.³⁹
   - Self-driving cars: AI systems that use sensors, cameras, and deep learning algorithms to navigate roads autonomously.⁴⁰

Narrow AI is currently the most common form of AI which we interact with daily, and it excels in specific, well-defined tasks but lacks the ability to adapt to new, unforeseen challenges.[41]

1. **General AI (Strong AI)**

   General AI refers to systems that possess generalized human intelligence.[42] These machines would be able to perform any intellectual task a human can, including learning, reasoning, and adapting to new environments without needing additional programming.[43] General AI remains largely theoretical, as no current AI system exhibits this level of flexibility.[44]

Example: While no true General AI exists today, the aspiration for such systems has been captured in various works of fiction, including HAL 9000 from 2001: A Space Odyssey[45] and the Replicants in Blade Runner.[46]

Building General AI would require a machine to possess consciousness, abstract reasoning, and emotional understanding—capabilities far beyond the current state of the field.[47]

1. **Super AI**

Super AI is a hypothetical form of intelligence that surpasses human intelligence in all aspects, including creativity, problem-solving, and decision-making.[48] Such an AI could potentially lead to the "technological singularity," where machines improve themselves autonomously and exponentially, leading to unforeseen and potentially transformative consequences for humanity.[49]

Example: The idea of Super AI is still speculative, though thinkers like Nick Bostrom have explored its implications in his book Superintelligence: Paths, Dangers, Strategies.[50]

Applications of AI Across Industries

AI is transforming industries across the board, from healthcare and finance to entertainment and logistics.[51]

- Healthcare: AI is used to analyze medical images, predict patient outcomes, and even assist in drug discovery.[52] IBM Watson Health has developed AI systems capable of suggesting treatment options for cancer patients by analyzing medical data.[53]
- Finance: AI-driven algorithms are employed in high-frequency trading, fraud detection, and risk management.[54] Robo-advisors, like Wealthfront and Betterment, provide personalized financial advice using machine learning models.[55]
- Entertainment: In video games, AI is used to create non-playable characters (NPCs) that adapt to the player's strategies.[56] Streaming services like Netflix employ AI to personalize recommendations.
- Transportation: Companies like Tesla, Waymo, and Uber are pioneering autonomous vehicles that rely on a combination of AI techniques, including computer vision and sensor fusion, to navigate environments.

In the words of Stephen Hawking, "Success in creating AI would be the biggest event in human history. Unfortunately, it might also be the last, unless we learn how to avoid the risks."[57]

Footnotes

1. Russell, S., & Norvig, P. (2020). *Artificial Intelligence: A Modern Approach*. Pearson.
2. McCarthy, J., Minsky, M. L., Rochester, N., & Shannon, C. E. (1955). A proposal for the Dartmouth summer research project on artificial intelligence.
3. Goodfellow, I., Bengio, Y., & Courville, A. (2016). *Deep Learning*. MIT Press.
4. Copeland, B. J. (2004). *The Story of Artificial Intelligence*. Oxford University Press.
5. Descartes, R. (1641). *Meditations on First Philosophy*.
6. Turing, A. M. (1950). Computing Machinery and Intelligence. *Mind*, 59(236), 433-460.
7. Ibid.
8. Ibid.
9. Turing, A. M. (1950). *Computing Machinery and Intelligence*. Mind, 59(236), 433-460.
10. McCarthy, J. (1956). *Proposal for the Dartmouth Summer Research Project on Artificial Intelligence*.
11. McCarthy, J., Minsky, M. L., Rochester, N., & Shannon, C. E. (1955). *A Proposal for the Dartmouth Summer Research Project on Artificial Intelligence*.
12. Newell, A., & Simon, H. A. (1956). *The Logic Theory Machine: A Complex Information Processing System*.
13. Newell, A., & Simon, H. A. (1956). *The Logic Theory Machine: A Complex Information Processing System*.
14. Ibid.
15. Crevier, D. (1993). *AI: The Tumultuous History of the Search for Artificial Intelligence*. BasicBooks.
16. Ibid.
17. Russell, S., & Norvig, P. (2020). *Artificial Intelligence: A

Modern Approach. Pearson.
18. Mitchell, T. M. (1997). *Machine Learning*. McGraw-Hill.
19. Bishop, C. M. (2006). *Pattern Recognition and Machine Learning*. Springer.
20. Quinlan, J. R. (1986). Induction of decision trees. *Machine Learning*, 1(1), 81-106.
21. Breiman, L. (2001). *Random Forests*. Machine Learning, 45(1), 5-32; Cortes, C., & Vapnik, V. (1995). *Support-Vector Networks*. Machine Learning, 20(3), 273-297.
22. McCulloch, W. S., & Pitts, W. (1943). A logical calculus of the ideas immanent in nervous activity. *The Bulletin of Mathematical Biophysics*, 5(4), 115-133.
23. Rumelhart, D. E., Hinton, G. E., & Williams, R. J. (1986). Learning representations by back-propagating errors. *Nature*, 323(6088), 533-536.
24. Hinton, G. E., Osindero, S., & Teh, Y. W. (2006). A fast-learning algorithm for deep belief nets. *Neural Computation*, 18(7), 1527-1554.
25. LeCun, Y., Bengio, Y., & Hinton, G. (2015). Deep learning. *Nature*, 521(7553), 436-444.
26. Krizhevsky, A., Sutskever, I., & Hinton, G. E. (2012). ImageNet classification with deep convolutional neural networks. *Advances in Neural Information Processing Systems*, 25, 1097-1105.
27. Schmidhuber, J. (2015). Deep learning in neural networks: An overview. *Neural Networks*, 61, 85-117.
28. Ibid.
29. Ibid.
30. Krizhevsky, A., Sutskever, I., & Hinton, G. E. (2012). *ImageNet Classification with Deep Convolutional Neural Networks*. Advances in Neural Information Processing Systems, 25.

31. Goodfellow, I., Bengio, Y., & Courville, A. (2016). *Deep Learning*. MIT Press.
32. Brown, T. et al. "Language Models are Few-Shot Learners." *Advances in Neural Information Processing Systems*, 2020. This paper introduces GPT-3, a large-scale language model capable of few-shot learning, marking a significant milestone in natural language processing with 175 billion parameters.
33. Silver, D. et al. "Mastering the Game of Go with Deep Neural Networks and Tree Search." *Nature* 529, 484–489 (2016). AlphaGo's success against world champion Fan Hui demonstrated the unprecedented strategic capabilities of AI in complex games, showcasing the synergy between reinforcement learning and neural networks.
34. LeCun, Y., Bengio, Y., & Hinton, G. "Deep Learning." *Nature* 521, 436–444 (2015). This foundational paper discusses how deep learning architectures—through multiple layers—enable breakthroughs in fields like computer vision, speech recognition, and game-playing AI.
35. Kaplan, A., & Haenlein, M. "Siri, Siri, in My Hand: Who's the Fairest in the Land? On the Interpretations, Illustrations, and Implications of Artificial Intelligence." *Business Horizons* 62, no. 1 (2019): 15-25Silver, D., et al. (2016). Mastering the game of Go with deep neural networks and tree search. *Nature*, 529(7587), 484-489.
36. Ibid.
37. Russell, S., & Norvig, P. (2020). *Artificial Intelligence: A Modern Approach*. Pearson. ↵
38. Ibid.
39. Ibid.
40. Hoy, M. B. (2018). Alexa, Siri, Cortana, and More: An Introduction to Voice Assistants. *Medical Reference Services Quarterly*, 37(1), 81-88.

41. Gomez-Uribe, C. A., & Hunt, N. (2015). The Netflix Recommender System: Algorithms, Business Value, and Innovation. *ACM Transactions on Management Information Systems*, 6(4), 13.
42. Litman, T. (2020). Autonomous Vehicle Implementation Predictions: Implications for Transport Planning. *Transportation Research Part A: Policy and Practice*, 132, 260-274.
43. Ibid.
44. Goertzel, B., & Pennachin, C. (Eds.). (2007). *Artificial General Intelligence*. Springer.
45. Ibid.
46. Russell, S., & Norvig, P. (2020). *Artificial Intelligence: A Modern Approach*. Pearson.
47. Kubrick, S., & Clarke, A. (1968). *2001: A Space Odyssey*. New American Library.
48. Scott, R. (1982). *Blade Runner*. Warner Bros.
49. Searle, J. R. (1980). Minds, brains, and programs. *Behavioral and Brain Sciences*, 3(3), 417-424.
50. Bostrom, N. (2014). *Superintelligence: Paths, Dangers, Strategies*. Oxford University Press.
51. Ibid.
52. Bostrom, N. (2014). *Superintelligence: Paths, Dangers, Strategies*. Oxford University Press.
53. Manyika, J., et al. (2017). *A Future that Works: Automation, Employment, and Productivity*. McKinsey Global Institute.
54. Topol, E. (2019). *Deep Medicine: How Artificial Intelligence Can Make Healthcare Human Again*. Basic Books.
55. IBM Watson Health. (2023). *IBM Watson Health: AI Solutions for Healthcare*. ↵
56. Smith, M. (2018). How Netflix uses AI and machine learning to make your favorite shows. *VentureBeat*.

57. Hawking, S. (2014). *Brief Answers to the Big Questions*. Bantam Books.

Chapter 2: The Intersection of AI and Faith

Several years ago, a colleague of mine posted on an internal chat group that he had written a sermon using AI. I found this to be both intriguing and theologically perplexing. I later turned my curiosity in Religion and AI into my second-day Rosh Hashanah talk in 2024. This chapter is the outcome of such practical rabbinic interactions.

In recent decades, artificial intelligence (AI) has moved from being a futuristic concept to an integral part of everyday life, influencing industries from healthcare and education to entertainment and transportation. With its capacity for rapid problem-solving, data analysis, and automation, AI has ushered in a new era of technological progress. But as with any transformative innovation, its growing ubiquity raises deep ethical and philosophical questions. These questions, while relevant to technologists and policymakers, also invite theological reflection. This chapter explores why it is essential to examine AI through a theological lens, and it introduces the central themes of this book, which is rooted in the conservative Jewish tradition and my own rabbinical training at the Schechter Institute for Judaic Studies in Jerusalem, yet it speaks to broader concerns about technology and faith.

Humanity has long sought to understand its place in the universe, and religious traditions provide frameworks for interpreting the world, our responsibilities, and the divine. Today, AI challenges some of these long-standing beliefs, such as the nature of intelligence, moral agency, and the relationship between the human and the divine. Grounded in Jewish teachings, but extending into universal religious and secular viewpoints, this chapter offers a comprehensive look at the pressing need to engage with AI's ethical dimensions, while simultaneously addressing broader theological issues in the context of rapid technological advancement.

Why Theological Inquiry into AI is Necessary

Theological inquiry into AI is not merely academic or speculative—it is urgent and necessary. As AI systems become more advanced, their ability to make autonomous decisions without direct human intervention compels us to question foundational concepts of personhood, ethics, and responsibility. Is there a point at which an AI, designed to think and act like a human, becomes something more than a tool? Can we consider it a moral agent, capable of ethical behavior, or is it forever bound by its programming? These questions, rooted in ethics and philosophy, intersect directly with theology, particularly concerning human nature, free will, and divine purpose.

Reconsidering Humanity and Creation

Central to the Jewish understanding of humanity is the concept of *Imago Dei*—the belief that humans are created in the image of God.[1] According to the Torah, this divine image sets humans apart from all other creatures, bestowing upon them unique capacities for moral reasoning, creativity, and self-determination. But what happens when machines begin to exhibit capabilities that rival or even surpass human intelligence? Are they mere extensions of human ingenuity, or do they reflect something more profound about creation itself?

AI forces us to grapple with the boundaries between humanity and technology. In many ways, the development of AI mirrors the role humans are assigned in Genesis as co-creators alongside God in the ever-unfolding realm of revelation. The act of building something with intellect and agency reflects the divine mandate to "fill the earth and subdue it" (Genesis 1:28).[2] Yet, this same narrative calls for humility, as humans are ultimately bound by the moral and ethical parameters set by God.

From the Jewish perspective, the development of AI can be seen as part of humanity's continuous quest to harness the world's potential.

However, Jewish teachings also caution against hubris. The story of the Tower of Babel, where humans attempted to build a structure to the heavens in defiance of God, serves as a timeless warning about the dangers of overreach.[3] It is possible to view AI as a modern "Tower of Babel," (Genesis 11: 1-9) wherein the quest for greater and greater intelligence risks displacing the very boundaries of what it means to be human.

The Ethical Challenge of AI: Moral Agency and Accountability

One of the most pressing concerns about AI is its capacity to make decisions without human oversight. From autonomous vehicles to AI-driven medical diagnoses, the possibility of machines taking on responsibilities traditionally held by humans raises critical questions about accountability and ethics. Who is to be held accountable when an AI system makes a mistake? Is it the creator, the user, or the system itself?

Judaism places immense value on the concept of *bechirah chofshit* (free will), which underpins human accountability.[4] According to Jewish tradition, moral agency is inextricably linked to the ability to choose between good and evil. Humans are responsible for their choices, and they are judged based on their capacity to act morally.[5] AI systems, however, operate on the basis of data and algorithms, devoid of the subjective moral experience that characterizes human decision-making. This absence of moral agency challenges traditional theological frameworks. Especially when it comes to tshuvah.

At the same time, Jewish teachings about moral responsibility emphasize the importance of ethical oversight and stewardship. In Deuteronomy 30:19, God presents humanity with a clear choice between life and death, blessing and curse, and exhorts them to "choose life".[6] In the context of AI, this directive calls on humans to ensure that their technological creations are used ethically and to prevent harm.

AI must be shaped by human values, not allowed to exist in an ethical vacuum. For the ramifications can be catastrophic.

This need for ethical oversight resonates with secular discussions on AI. For instance, contemporary AI ethics frameworks, such as Asimov's famous Three Laws of Robotics,[7] advocate for the primacy of human safety and well-being in AI decision-making. While these guidelines serve as useful starting points, they lack the depth of moral reasoning present in religious traditions, which ground human action in an understanding of divine will and moral law.[8]

Jewish Perspectives on Ethics and Technology

Judaism has always sought to balance the demands of modernity with the timeless wisdom of Torah and halacha (Jewish law). The intersection of ethics and technology is not a new concern for Jewish thinkers. Historically, the development of new technologies has prompted Jewish legal authorities to issue *responsa*—halachic rulings that apply Jewish law to contemporary challenges.

Halachic Approaches to New Technologies

In the case of AI, there are several halachic considerations. First, Jewish law places a high value on privacy and the protection of personal information.[9] As AI systems become more adept at collecting and analyzing personal data, halachic authorities must address how Jewish law applies to issues like data privacy and surveillance. The Talmudic principle of *hezek re'iyah*—the prohibition against causing harm by seeing or intruding on another's privacy—may provide a foundation for contemporary discussions on AI-driven surveillance.[10]

Second, Jewish ethics emphasizes the dignity of work and the importance of human labor.[11] With AI automating tasks that were once the exclusive domain of human workers, questions arise about how these developments align with Jewish teachings on labor and economic justice. The Torah's emphasis on fair treatment of workers and the prohibition against exploitation may guide the ethical use of AI in industries where automation could lead to job displacement.[12]

Maintaining Human Agency in the Age of AI

While AI promises to enhance human capabilities in remarkable ways, Jewish teachings warn against surrendering human agency to technology. The Jewish concept of partnership with God in the ongoing process of creation assumes that humans remain active participants in ethical decision-making. In the age of AI, this means that humans must retain ultimate responsibility for the systems they create and deploy.

We learn in the Mishnah that "everything is foreseen, yet free will is given" (Pirkei Avot 3:15).[13] This paradox, which lies at the heart of Jewish theology, suggests that while God has knowledge of all future events, humans still possess the freedom to make moral choices. In a world where AI systems can predict outcomes and suggest actions based on vast amounts of data, the human capacity for moral deliberation remains essential. AI may assist in decision-making, but it should not replace the human role in ethical reasoning.

The Nature of the Divine in the Age of AI

Theological inquiry into AI ultimately touches on fundamental questions about the nature of the divine and humanity's relationship with God. Many religious traditions hold that humans possess a soul, a divine spark that connects them to God and endows them with unique moral and spiritual capacities. But if AI can mimic human cognition, does it challenge the traditional understanding of the soul?

The Soul and Consciousness

In Jewish thought, the soul is seen as the seat of consciousness and moral awareness, a gift from God that distinguishes humans from all other forms of life.[14] The soul is not merely an emergent property of physical processes, as some secular thinkers suggest, but a reflection of the divine presence within each individual. AI, no matter how advanced, lacks this divine spark, and thus cannot possess the same moral or spiritual capacities as humans.[15]

However, AI does raise intriguing questions about consciousness. While most current AI systems operate without consciousness, some philosophers and technologists speculate about the possibility of creating machines that possess self-awareness. If this were to happen, it would challenge theological conceptions of the soul. From a Jewish perspective, even if machines were to achieve a form of consciousness, they would still lack the divine essence that defines human beings.[16] The soul is not something that can be engineered or replicated through technological means—it remains a unique gift from God.

AI and Divine Providence

Jewish tradition teaches that God is actively involved in the world, guiding events according to a divine plan[17]. This belief in divine providence raises questions about the role of technology in the unfolding of God's plan. Is AI part of the divine blueprint, or does it represent a deviation from the intended course of creation?

Rabbi Abraham Joshua Heschel argued that technological progress, like all human endeavors, must be tempered by a sense of awe and reverence for the divine.[17] For Heschel, the danger of modern technology lies not in its capabilities, but in humanity's tendency to see itself as the master of the universe, independent of God. AI, with its promise of near-limitless knowledge and control, can easily lead to this kind of hubris. Jewish theology offers a counterbalance to this tendency by reminding us that all knowledge and power ultimately come from God.

At the same time, Jewish thought also sees human creativity and technological innovation as part of the divine plan. The Talmud teaches that "the Holy One, blessed be He, created the world to be perfected" (Bereshit Rabbah 11:6).[18] This suggests that technological advancements, including AI, can be viewed as part of humanity's divinely ordained mission to improve the world. However, this responsibility must be exercised with caution, ensuring that technological progress aligns with ethical and spiritual values.

Bridging Religious and Secular Perspectives

While this book is written from a Jewish theological perspective, the issues it raises have universal relevance. AI's impact on human identity, ethics, and spirituality is a concern shared by people of all faiths and none. By engaging in dialogue with both religious and secular viewpoints, we can deepen our understanding of AI's ethical and theological implications.

Comparative Religious Perspectives on AI

Different religious traditions offer diverse perspectives on AI and its ethical challenges. In Christianity, for example, the concept of stewardship emphasizes the responsible use of technology for the common good.[19] Christian theologians have raised concerns about AI's potential to dehumanize individuals by reducing them to mere data points, urging a return to the Gospel's message of human dignity.[20]

Academic Islamic scholars, too, have explored the ethical dimensions of AI, emphasizing the importance of justice and accountability in its deployment.[21] Islamic teachings on the sanctity of life and the pursuit of knowledge encourage the development of technologies that benefit humanity while adhering to ethical principles.[22]

By engaging with these perspectives, we can develop a more comprehensive ethical framework for AI, one that is informed by a wide range of religious and moral traditions.

Secular Ethical Frameworks and AI

Secular ethics also offers valuable insights into the challenges posed by AI. Philosophers such as Immanuel Kant and John Stuart Mill have provided foundational theories of ethics that continue to influence contemporary discussions on AI.[23] Utilitarianism, for example, advocates for maximizing overall happiness and well-being, a principle that can guide the ethical development of AI systems.[24] Similarly, Kant's deontological ethics, with its emphasis on treating individuals as ends in themselves rather than means to an end, provides a useful counterpoint to the potential dehumanization of AI.[25]

While these secular ethical frameworks are important, they often lack the spiritual and existential dimensions that religion provides. By integrating religious and secular perspectives, we can create a more holistic approach to AI ethics, one that addresses both the material and spiritual needs of humanity.

Conclusion

In my opinion, the intersection of AI and faith represents one of the most profound ethical and theological challenges of our time. As AI continues to reshape society, it is essential to engage with the ethical, philosophical, and theological questions it raises. From a Jewish perspective, the development of AI invites us to reflect on the nature of humanity, the role of free will, and our responsibility to use technology ethically. At the same time, these questions resonate across religious and secular traditions, emphasizing the universal importance of thoughtful engagement with AI.

In this chapter, we have introduced the central themes and the object of this book: the need for a theological inquiry into AI, the challenges AI poses to traditional concepts of morality and agency, and the broader implications of AI for our understanding of the divine. As we move forward, we will continue to explore these themes, drawing on Jewish sources and engaging with a wide range of religious and secular perspectives. Ultimately, our goal is to provide a framework for thinking about AI that is both ethically sound and spiritually meaningful.

Footnotes

1. Genesis 1:27, Torah.
2. Ibid.
3. Genesis 11:1-9, Torah.
4. Rabbi Samson Raphael Hirsch, *The Nineteen Letters*, 1834.
5. Maimonides, *Hilchot Teshuva* 5:1.
6. Deuteronomy 30:19, Torah.
7. Isaac Asimov, *Runaround*, 1942.
8. *Jewish Law and Modern Ethics*, David Shatz, 2002.
9. *Data Privacy in Jewish Law*, Rebecca Weiner, 2023.
10. Talmud, Tractate Bava Batra 2a.
11. *Jewish Views on Labor and Economy*, Nechama Leibowitz, 1994.
12. Ibid.
13. Pirkei Avot 3:15, Talmud.
14. *The Soul in Jewish Thought*, Rachel Adler, 2022.
15. Ibid.
16. *AI and the Quest for Consciousness*, Michael Levin, 2021.
17. Rabbi Abraham Joshua Heschel, *The Sabbath*, 1951.
18. *Bereshit Rabbah* 11:6, Midrash.
19. *Christian Stewardship and Technology*, Paul Johnson, 2022.
20. *Christian Ethics and Artificial Intelligence*, Laura Smith, 2021.
21. *Islamic Perspectives on Artificial Intelligence*, Fatima Zahra, 2023.
22. *Islamic Techno-Ethics*, Ayesha Abdullah, 2023.
23. *Ethical Theories in the Age of AI*, Thomas Green, 2022.
24. John Stuart Mill, *Utilitarianism*, 1863.
25. Immanuel Kant, *Groundwork of the Metaphysics of Morals*, 1785.

Chapter 3: Theological Foundations: Divine Creation and Human Innovation

In the rapidly evolving landscape of the 21st century, where technological advancements continuously redefine the boundaries of possibility, the intersection of artificial intelligence (AI) and religion invites profound reflection. This chapter delves into the theological underpinnings of human creativity as articulated in Jewish thought, particularly through the lens of the concept of *B'tzelem Elohim*, or in the "Image of God." By examining how this foundational belief informs our understanding of human innovation and responsibility, we aim to illuminate the moral imperatives that arise when humanity engages in the creation of intelligent machines. Furthermore, we will explore parallels with other religious traditions to provide a comprehensive view of creation, stewardship, and the ethical challenges posed by our technological endeavors.

B'tzelem Elohim: The Image of God and Human Creativity

The concept of *B'tzelem Elohim*, derived from Genesis 1:26–27, holds a central place in Jewish theology, positing that humans are created in the divine image. This profound assertion not only establishes the inherent dignity of every individual but also articulates a theological framework for understanding human creativity as a divine attribute. The Talmud expands upon this idea, emphasizing the unique role of humanity in the divine plan and affirming that every person reflects aspects of God's essence.[1] This notion of divine likeness implies that our creative capacities are not merely utilitarian but are imbued with holiness and spiritual significance. Two cornerstones of the covenantal relationship between God and man.

The late Chief Rabbi of the United Kingdom, Rabbi Jonathan Sacks Z"L, asserts that to be human is to engage in creativity, to take the raw materials of the world and reshape them in ways that resonate with higher purposes.² This creative drive manifests through various forms of expression—art, science, technology—and can be seen as a reflection of the divine creative act. Our technological endeavors, including the development of AI, can thus be understood as an extension of our spiritual calling to innovate, improve the world, and make both more accessible to all.

However, with this creative power comes immense responsibility. As we engage in the creation of intelligent machines, we must grapple with questions of ethics, purpose, and the implications of what it means to be human. Rabbi Sacks cautions against the notion that technology is inherently good or evil; rather, it is how we utilize our innovations that ultimately reflects our ethical and moral values.³ The challenge lies in ensuring that our technological advancements honor the dignity inherent in *B'tzelem Elohim*.

The Distinction Between Divine and Human Creation

To navigate the theological implications of our technological pursuits, we must address the distinction between divine creation and human innovation. Theologians have long contemplated the nature of creation, with divine acts characterized as fundamentally good and purposeful. In contrast, human inventions—including AI—can yield mixed results, depending on the *yetzer* (motivations) and choices of their creators. This ambiguity necessitates a rigorous ethical framework that aligns with the moral imperatives outlined in religious texts.

In Jewish thought, the concept of *Tikkun Olam*, or "repairing the world," serves as a guiding principle for human creativity.⁴ This mandate compels us to engage in actions that contribute positively to society and the environment. The development of AI offers the potential for significant advancements in areas such as healthcare, education, and sustainability. Yet, it also presents the risk of exacerbating societal inequalities, diminishing personal agency, and even compromising moral standards. Thus, the call to *Tikkun Olam* must inform our approach to technology, urging us to prioritize ethical considerations in our innovations.

The philosophical discourse surrounding creativity extends beyond the Jewish tradition. In Christian theology, particularly within the writings of figures like Augustine and Aquinas, there is an emphasis on the idea that human creativity is a participation in God's own creative act. Augustine writes, "God created humans in order that they might create,"[5] suggesting a divine endorsement of human innovation. This sentiment echoes through various theological frameworks, reinforcing the idea that our capacity to create—whether through art, technology, or AI—reflects our divine origin.

The Jewish principle of *Ba'al Tashchit* teaches us to avoid waste and destruction, encouraging a relationship of stewardship toward creation.[6] This principle becomes especially pertinent in our dealings with technology. The development of AI, when approached with the mindset of stewardship, can foster innovations that align with the values of preservation and enhancement of life. For instance, AI can optimize resource management, reduce waste, and improve efficiency in various sectors, from agriculture to urban planning. Applications in precision agriculture, for example, utilize AI to minimize water use and pesticide application, thereby conserving resources and reducing environmental impact.

However, without a commitment to stewardship, technological advancements can lead to environmental degradation and ethical lapses, undermining the very principles we are called to uphold. Consider the environmental consequences of data centers that power AI systems; their energy consumption can be staggering, contributing to climate change if not managed responsibly. A commitment to *Ba'al Tashchit* necessitates that we assess the ecological impact of our technological creations, advocating for sustainable practices that honor the divine mandate of stewardship.

Agency, Free Will, and Moral Responsibility

The emergence of intelligent machines raises critical questions about Free Will and moral responsibility. Central to Jewish theology is the belief in free will, as articulated in texts in Deuteronomy, where individuals are urged to choose life and good over death and evil.[7] This notion of free will is foundational to our understanding of morality and accountability. However, as AI systems become more autonomous, we must ask: how do we maintain human agency in the face of machines capable of making decisions on our behalf?

Rabbi David Hartman, founder of the Shalom Hartman Institute in Jerusalem, has articulated that moral responsibility cannot be outsourced to technology. He asserts that, "Moral decisions require the engagement of human consciousness."[8] as we develop increasingly sophisticated AI, we must ensure that human judgment remains at the forefront, guiding the ethical implications of technological innovations.

Jewish ethics places significant weight on *kavanah*, or intention behind actions. The Mishnah teaches that the "heart is the seat of moral decision-making,"[9] indicating that the motivations driving our technological pursuits are as critical as the outcomes.

Footnotes

1. Babylonian Talmud, Sanhedrin 37a.
2. Jonathan Sacks, *The Home We Build Together: Recreating Society* (London: Continuum, 2007).
3. Jonathan Sacks, *To Heal a Fractured World: The Ethics of Responsibility* (New York: Schocken, 2005).
4. Pirkei Avot 2:21.
5. Augustine, *On Free Choice of the Will*, trans. Thomas Williams (Indianapolis: Hackett, 1993).
6. Mishnah, Shabbat 7:2.
7. Deuteronomy 30:19.
8. David Hartman, *A Living Covenant: The Innovative Spirit in Traditional Judaism* (New York: Free Press, 1998).

Chapter 4: Ethical AI: Bridging Tradition and Technology

In the rapidly evolving landscape of artificial intelligence (AI), the convergence of technology and theology presents a unique opportunity to scrutinize the ethical underpinnings of AI development and its societal implications. As a conservative rabbi and AI ethicist, my interest and goal is to harmonize traditional Jewish ethical principles with contemporary technological advancements. This chapter explores how foundational Jewish values, particularly Tikkun Olam (repairing the world), can inform and guide the ethical creation and deployment of AI technologies. Furthermore, by integrating ethical guidelines from a spectrum of multi-faceted religious traditions and secular philosophies, this discourse aims to present a comprehensive and multifaceted perspective on AI ethics. The objective is to delineate practical implications and offer actionable guidance on how these ethical frameworks can shape AI policies and practices, ensuring that technology serves the greater good while respecting our deepest moral and spiritual values.

Jewish Ethical Principles and AI
Tikkun Olam: Repairing the World

Tikkun Olam, a cornerstone of Jewish ethical thought, encapsulates the responsibility to contribute to the betterment of society and the world at large. One of the ways that we accomplish this is through Tikkun Atzmi (repairing oneself) and taking responsibility for our wider surroundings. In the realm of AI, this principle serves as a moral compass, directing the development of technologies that aim to alleviate human suffering, promote justice, and enhance overall well-being.[2] The application of Tikkun Olam in AI is multifaceted, encompassing efforts to bridge societal gaps, enhance accessibility, and foster sustainable practices.

Alleviating Suffering Through AI

AI has the potential to significantly reduce human suffering by addressing critical issues such as healthcare disparities and access to education. For example, AI-driven diagnostic tools can democratize healthcare by providing accurate and timely medical assessments in underserved regions, thereby embodying the Tikkun Olam imperative to rectify societal imbalances.[3] Similarly, AI-powered educational platforms can offer personalized learning experiences, ensuring that quality education is accessible to all, regardless of geographical or socio-economic barriers.[4]

Promoting Justice and Equity

Tikkun Olam also emphasizes the pursuit of justice and equity. AI systems can be designed to identify and mitigate biases that perpetuate social injustices. By employing machine learning algorithms that are trained on diverse and representative datasets, developers can minimize discriminatory outcomes in areas such as hiring, lending, and law enforcement.[5] Moreover, AI can facilitate more equitable resource distribution by optimizing supply chains and enhancing the efficiency of humanitarian aid delivery.[6] All this adds up to wider universal trust within divergent segments.

Enhancing Environmental Stewardship

Environmental degradation poses one of the most pressing challenges of our time. AI applications in environmental monitoring and climate modeling can provide invaluable insights into climate change patterns, enabling more effective and timelier responses.[7] For instance, AI-driven systems can optimize energy consumption in smart grids, reduce waste through predictive maintenance in manufacturing, and enhance the precision of agricultural practices, thereby promoting sustainable stewardship of the Earth in alignment with Jewish environmental ethics.[8]

Pikuach Nefesh: The Primacy of Life

Pikuach Nefesh, the principle that the preservation of human life overrides almost all other commandments (except for murder, incest, and idol worship), is profoundly relevant in the context of AI ethics.[9] This principle mandates that AI technologies prioritize human safety and well-being above all else, ensuring that advancements do not compromise the sanctity of life.

Ensuring AI Safety and Reliability

AI systems, particularly those deployed in critical areas such as healthcare, transportation, and security, must undergo rigorous testing and validation to ensure their reliability and safety.[10] Implementing fail-safe mechanisms, redundancy protocols, and continuous monitoring can prevent unintended consequences and mitigate risks associated with AI failures.[11] Additionally, transparency in AI decision-making processes fosters trust and allows for accountability in cases where AI systems may impact human lives.[12]

Ethical Decision-Making in Autonomous Systems

The deployment of autonomous systems, such as self-driving cars and autonomous drones, raises complex ethical questions about decision-making in life-and-death scenarios.[13] Adhering to Pikuach Nefesh requires that these systems are programmed with ethical frameworks that prioritize human life and minimize harm. This involves not only technical safeguards but also ongoing ethical evaluations to adapt to evolving societal norms and values.[14] The ethical dilemma between safety and national defense is an issue currently being debated due to the ongoing wars in both Ukraine and the Middle East.

Dina d'Malkhuta Dina: The Law of the Land

The principle of Dina d'Malkhuta Dina, meaning "the law of the land is the law," underscores the importance of adhering to the legal frameworks and societal norms established by governing authorities.[15] In the realm of AI, this principle advocates for compliance with laws related to data privacy, intellectual property, and ethical standards, ensuring that AI development aligns with both legal requirements and ethical expectations.[16]

Navigating Legal Compliance in AI Development

AI developers must navigate a complex landscape of regulations that govern data usage, privacy, and security. Adhering to laws such as the General Data Protection Regulation (GDPR) in Europe or the California Consumer Privacy Act (CCPA) in the United States ensures that AI systems respect individual privacy rights and data protection standards.[17] Compliance not only fosters trust among users but also mitigates legal risks associated with data breaches and misuse.[18]

Collaborative Policy-Making

The intersection of AI and law necessitates collaboration between technologists, ethicists, religious communities, and policymakers. By engaging with diverse stakeholders, AI regulations can be crafted to reflect a broad spectrum of ethical values and societal needs, ensuring that technological advancements are both legally sound and ethically responsible.[19] This collaborative approach promotes the creation of policies that balance innovation with necessary safeguards, fostering an environment where ethical AI can flourish.[20]

Comparative Ethical Frameworks

Christian Ethics and Compassionate AI Applications

Christian ethical teachings, particularly the concept of Agape (selfless love), resonate deeply with the ideals of ethical AI development. Agape emphasizes the well-being and dignity of others, advocating for actions that promote compassion, altruism, and the common good.[21]

Incorporating Agape into AI development encourages the creation of technologies that prioritize human welfare. For instance, AI applications in mental health can provide support and resources to individuals in need, reflecting the Christian call to care for the vulnerable and marginalized.[22] Additionally, AI-driven initiatives in disaster response and recovery embody the spirit of Agape by facilitating swift and effective humanitarian aid.[23]

Islamic Ethics and Promoting Justice and Fairness

Islamic ethics, grounded in the principles of Adl (justice) and Ihsan (excellence), offer valuable insights into equitable and high-quality AI development.[24] These principles mandate that AI systems operate fairly, without bias, and strive for excellence in their performance and impact.[25]

Adl compels AI developers to ensure that their systems do not perpetuate or exacerbate existing injustices. This involves actively identifying and mitigating biases in AI algorithms, ensuring that outcomes are equitable across different demographic groups.[26] Furthermore, Ihsan encourages the pursuit of excellence in AI performance, fostering technologies that are not only effective but also ethically sound and socially beneficial.[27]

Islamic ethics also emphasize accountability and transparency in all actions, including technological development. Ensuring that AI systems are transparent in their decision-making processes and that developers are accountable for the ethical implications of their creations aligns with the Islamic imperative to uphold moral integrity.[28]

Secular Philosophies
Utilitarianism and Deontological Ethics

Utilitarianism, as articulated by philosophers like Jeremy Bentham and John Stuart Mill, advocates for actions that maximize overall happiness and minimize suffering.[29] In the context of AI, this translates to designing systems that produce the greatest good for the greatest number. For example, AI applications in public health can optimize resource allocation, ensuring that interventions reach those most in need and thereby enhancing collective well-being.[30]

Deontological ethics, particularly Immanuel Kant's categorical imperative, emphasizes the importance of adhering to moral rules and duties regardless of the consequences.[31] This framework ensures that

AI respects individual rights and autonomy, preventing technologies from being used in ways that violate fundamental ethical principles. For instance, AI systems must respect user privacy and consent, treating individuals as ends in themselves rather than a mere means to an end.[32]

Interfaith Dialogue on AI Ethics

Incorporating ethical guidelines from diverse religious traditions fosters a richer, more inclusive approach to AI ethics.[33] Interfaith dialogue bridges gaps between different moral perspectives, promoting a holistic understanding of ethical AI that transcends cultural and religious boundaries.[34] This inclusive approach ensures that AI development is sensitive to a wide array of values and principles, enhancing its ethical robustness and societal acceptance.[35]

Furthermore, interfaith collaborations can lead to the establishment of universal ethical standards for AI that draw upon the strengths of various religious and philosophical traditions.[36] By integrating diverse ethical insights, these standards can address complex moral dilemmas posed by AI technologies, ensuring that AI serves the common good across different cultural contexts.[37]

Engaging in interfaith dialogue fosters mutual understanding and respect among different religious communities, facilitating cooperative efforts in AI ethics.[38] This collaborative spirit can lead to the development of AI systems that are not only technically proficient but also culturally and ethically sensitive, promoting harmony and trust among diverse societal groups.[39]

Practical Implications for AI Development
Ethical AI Design

Integrating ethical principles into AI design is paramount for creating technologies that align with societal and moral standards.[40] This involves embedding values such as fairness, accountability, and transparency from the outset of the development process.[41] Ethical AI design requires a proactive approach, anticipating potential ethical challenges and addressing them through thoughtful and deliberate design choices.[42]

Fairness and Bias Mitigation

Ensuring fairness in AI involves identifying and mitigating biases that can lead to discriminatory outcomes.[43] This requires diverse and representative training data, as well as algorithms designed to detect and correct biases. Additionally, involving diverse teams in the AI development process can enhance the identification and resolution of potential biases, fostering more equitable technologies.[44]

Accountability and Transparency

Accountability in AI development means that developers and organizations are responsible for the ethical implications of their technologies.[45] This can be achieved through clear documentation of AI decision-making processes, transparency in data usage, and mechanisms for addressing grievances and ethical breaches.[46] Transparent AI systems enable users to understand how decisions are made, fostering trust and enabling accountability.

Sustainability and Long-Term Impact

Ethical AI design also considers the long-term impact of technologies on society and the environment. This involves assessing the sustainability of AI systems, minimizing their environmental footprint, and ensuring that they contribute positively to societal well-being over time. Sustainable AI practices align with the Jewish commitment to Tikkun Olam by promoting enduring solutions to global challenges.

Policy and Regulation

Crafting Ethical AI Legislation

Legislation aimed at regulating AI must reflect a nuanced understanding of ethical principles and societal needs. This includes establishing standards for data privacy, algorithmic transparency, and accountability, as well as creating frameworks for ethical AI certification and auditing. By embedding ethical considerations into legal frameworks, policymakers can ensure that AI technologies are developed and deployed in ways that respect human dignity and promote the common good.

International Cooperation and Standards

AI development is a global endeavor, necessitating international cooperation to establish universal ethical standards. Collaborative efforts among nations can lead to the creation of globally recognized guidelines that promote ethical AI practices, prevent harmful uses of AI, and ensure that the benefits of AI are equitably distributed. International standards also facilitate cross-border collaboration and the sharing of best practices, enhancing the overall ethical landscape of AI development.

Education and Awareness

Ethical Training in AI Curriculum

Integrating ethics into AI education ensures that future technologists are well-versed in the moral implications of their work. This can be achieved through dedicated courses on AI ethics, interdisciplinary programs that combine technology and humanities, and practical training on ethical decision-making and bias mitigation. By embedding ethics into the core of AI education, institutions can cultivate a generation of AI professionals committed to responsible and ethical innovation. This should be true in every theological program, irrespective of faith or denomination.

Public Awareness and Engagement

Raising public awareness about AI ethics is crucial for fostering informed and engaged communities. Public seminars, workshops, and informational campaigns can educate individuals about the ethical dimensions of AI, empowering them to advocate for responsible AI practices and policies. Engaged and informed citizens are better equipped to hold developers and policymakers accountable, ensuring that AI technologies align with societal values and ethical standards.

Community Engagement

Engaging with diverse communities ensures that AI technologies address the needs and values of different societal groups. Community input can guide the development of AI applications that are culturally sensitive and ethically sound, enhancing their acceptance and effectiveness. This inclusive approach fosters trust and collaboration between AI developers and the communities they serve, promoting technologies that are truly beneficial and respectful of diverse perspectives.

Participatory Design and Development

Participatory design actively involves community members in the AI development process, from conception to deployment. This collaborative approach ensures that AI systems are tailored to the specific needs and values of different communities, promoting inclusivity and relevance. By incorporating diverse voices, AI developers can create technologies that are more equitable, accessible, and responsive to the unique challenges faced by various societal groups.

Cultural Sensitivity and Respect

AI technologies must be designed with cultural sensitivity and respect for diverse traditions and values. This involves understanding and honoring the cultural contexts in which AI systems will be deployed, ensuring that they do not inadvertently disrespect or marginalize any group. Cultural sensitivity enhances the ethical robustness of AI, fostering technologies that are respectful, inclusive, and harmonious with the cultural fabric of society.

Ensuring AI Serves the Greater Good
Ethical Auditing

Regular ethical audits of AI systems are essential for identifying and mitigating potential harms, ensuring that technologies remain aligned with ethical principles. These audits involve systematic evaluations of AI systems to assess their impact on various aspects of society, including economic, social, and environmental dimensions.

Ethical auditing requires the development of comprehensive evaluation frameworks that encompass a wide range of ethical considerations. These frameworks should assess factors such as fairness, accountability, transparency, and sustainability, providing a holistic view of the AI system's ethical performance. By employing diverse evaluation criteria, auditors can ensure that AI technologies adhere to ethical standards and contribute positively to society.

Ethical auditing is not a one-time process but requires continuous monitoring and improvement. As AI technologies evolve, so do their ethical implications. Ongoing audits ensure that AI systems remain compliant with ethical guidelines and adapt to emerging challenges and societal changes. Continuous improvement processes foster a culture of ethical vigilance, ensuring that AI technologies remain responsible and beneficial over time.

Accountability Mechanisms

Establishing clear accountability mechanisms is vital for addressing ethical breaches in AI deployment. This involves defining responsibilities, implementing oversight structures, and enforcing consequences for unethical practices. Accountability ensures that individuals and organizations are held responsible for the ethical implications of their AI systems, fostering a culture of integrity and responsibility.

Clear definitions of roles and responsibilities are essential for effective accountability in AI development. This includes delineating the ethical obligations of developers, organizations, and stakeholders involved in the creation and deployment of AI technologies. By establishing clear lines of responsibility, accountability mechanisms ensure that ethical standards are upheld throughout the AI lifecycle.

Oversight Structures and Governance

Implementing robust oversight structures and governance frameworks is crucial for maintaining ethical standards in AI. Independent ethical review boards, regulatory bodies, and internal compliance teams can provide oversight and ensure that AI systems adhere to established ethical guidelines. These structures facilitate the identification and resolution of ethical issues, promoting responsible AI development and deployment.

Enforcement of Ethical Standards

Enforcing ethical standards requires the implementation of effective policies and consequences for non-compliance. This can include sanctions, penalties, and corrective measures for organizations and individuals that violate ethical guidelines. Effective enforcement deters unethical practices and reinforces the importance of ethical conduct in AI development.

Promoting Inclusivity

Ensuring that AI benefits are equitably distributed across different populations aligns with the ethical imperative to promote social justice. Inclusive AI development considers the diverse needs and perspectives of all societal members, preventing the exacerbation of existing inequalities. This commitment to inclusivity fosters technologies that are fair, accessible, and beneficial to all, contributing to the greater good.

Equitable Access to AI Technologies

Promoting equitable access to AI technologies ensures that all individuals and communities can benefit from advancements in AI. This involves addressing barriers such as cost, digital literacy, and infrastructure, enabling widespread access to AI-driven solutions. Equitable access fosters social inclusion and empowers marginalized communities, aligning with the ethical goal of promoting justice and equality.

Representation and Diversity in AI Development

Encouraging representation and diversity within AI development teams enhances the inclusivity and relevance of AI systems. Diverse teams bring varied perspectives and experiences, leading to more comprehensive and culturally sensitive AI solutions. By fostering diversity in AI development, we can create technologies that better reflect and serve the multifaceted nature of society.

Addressing Digital Divide and Technological Inequities

The digital divide, characterized by unequal access to technology, poses significant ethical challenges in AI development. Addressing these inequities requires concerted efforts to bridge the gap between different socio-economic groups, ensuring that AI advancements do not disproportionately benefit the privileged while neglecting the underserved. Initiatives such as affordable internet access, digital literacy programs, and inclusive technology policies are essential for mitigating the digital divide.

Conclusion

The convergence of technology and theology offers a unique and profound lens through which to examine the ethical dimensions of AI. By grounding AI development in foundational Jewish principles like Tikkun Olam and integrating ethical insights from a diverse array of religious and secular traditions, we can guide the creation of AI systems that uphold human dignity, promote justice, and contribute to the common good. This multifaceted approach ensures that AI technologies are not only technically proficient but also ethically robust and socially beneficial. As AI continues to evolve, it is imperative that ethical considerations remain at the forefront, fostering a symbiotic relationship between technology and our deepest moral and spiritual values. By doing so, we can harness the transformative potential of AI to repair and enhance the world, in harmony with our enduring commitment to ethical integrity, religious authenticity, and the greater good.

Footnotes

1. Maimonides, *Mishnah Torah*, Laws of Kings and Wars, 8:11. *Tikkun Olam* is a Hebrew phrase meaning "repairing the world," a fundamental concept in Jewish ethics emphasizing social justice and the responsibility to improve society.
2. Buber, *Tikkun Olam and the Pursuit of Justice*, Jewish Ethics Review, 2020. The application of AI in addressing global challenges aligns with *Tikkun Olam* by striving to create a more equitable and sustainable world.
3. Smith, "AI in Healthcare: Bridging the Gap," *Journal of Medical Ethics*, 2022. AI-driven diagnostic tools can reduce disparities in healthcare access by providing accurate and timely medical assessments in regions with limited resources.
4. Lee, "AI and Education: Expanding Access and Quality," *Educational Technology Quarterly*, 2023. AI-powered educational platforms can offer personalized learning experiences, ensuring quality education is accessible to all, regardless of geographical or socio-economic barriers.
5. Johnson, "Mitigating Bias in AI Systems," *AI Ethics Journal*, 2021. AI systems designed to identify and mitigate biases can help prevent discriminatory outcomes in areas such as hiring, lending, and law enforcement.
6. Goldstein, "AI for Humanitarian Aid," *International Development Review*, 2023. AI can facilitate more equitable resource distribution by optimizing supply chains and enhancing the efficiency of humanitarian aid delivery.
7. Thompson, "AI in Environmental Monitoring," *Climate Technology Review*, 2023. AI applications in environmental monitoring and climate modeling provide invaluable insights into climate change patterns, enabling more effective and timely responses.

8. Cohen, "Sustainable AI Practices," *Environmental Ethics and Technology*, 2024. AI-driven systems can optimize energy consumption in smart grids, reduce waste through predictive maintenance in manufacturing, and enhance the precision of agricultural practices.
9. *Shulchan Aruch*, Yoreh De'ah 157:1. *Pikuach Nefesh* dictates that saving a life overrides most other commandments, underscoring the paramount importance of human life in Jewish law.
10. Davis, "Ensuring AI Safety," *AI Safety Journal*, 2022. Rigorous testing and validation are essential to ensure AI systems' reliability and safety, especially in critical areas such as healthcare, transportation, and security.
11. Nguyen, "Fail-Safe Mechanisms in AI," *Technology and Society*, 2023. Implementing fail-safe mechanisms, redundancy protocols, and continuous monitoring can prevent unintended consequences and mitigate risks associated with AI failures.
12. Rodriguez, "Transparency in AI Decision-Making," *Journal of AI Ethics*, 2022. Transparency in AI decision-making processes fosters trust and accountability, allowing for the identification and resolution of ethical breaches.
13. Patel, "Ethical Dilemmas in Autonomous Systems," *Autonomous Technology Review*, 2023. The deployment of autonomous systems raises complex ethical questions about decision-making in life-and-death scenarios.
14. O'Connor, "Ethical Frameworks for Autonomous AI," *Journal of AI Ethics*, 2024. Adhering to *Pikuach Nefesh* requires programming autonomous systems with ethical frameworks that prioritize human life and minimize harm.
15. *Talmud*, Gittin 10a. *Dina d'Malkhuta Dina* reflects the Jewish obligation to respect and adhere to the laws of

governing authorities, promoting social order and legal compliance.
16. Kleinberg, "Legal and Ethical Considerations in AI," *Harvard Law Review*, 2022. Compliance with legal frameworks ensures that AI development respects societal norms and protects individual rights, fostering trust in technology.
17. European Commission, "Ethics Guidelines for Trustworthy AI," 2019. Adherence to data privacy laws like GDPR ensures that AI systems protect individual privacy rights and data security.
18. Kleinberg, "Legal and Ethical Considerations in AI," *Harvard Law Review*, 2022. Compliance with data privacy and security laws mitigates legal risks associated with data breaches and misuse.
19. Lee, "Policy-Making in AI: Incorporating Ethical Perspectives," *Policy Studies Journal*, 2022. Engaging with religious and philosophical communities in policy-making ensures that AI regulations reflect a broad spectrum of ethical values and societal needs.
20. Nguyen, "Balancing Innovation and Ethics in AI Regulation," *Regulatory Affairs Journal*, 2023. A collaborative approach to AI regulation helps balance innovation with necessary ethical safeguards, fostering trust and acceptance of AI technologies.
21. Augustine, *On Christian Doctrine*, Book III. *Agape* in Christian ethics emphasizes selfless love and the well-being of others, principles that can guide the compassionate use of AI technologies.
22. Brown, "Christian Ethics and AI," *Theology Today*, 2021. Compassionate AI applications, such as those in mental health, embody the Christian ethic of *Agape* by prioritizing human welfare.
23. Williams, "AI in Humanitarian Aid," *Disaster Response*

Journal, 2022. AI-driven initiatives in disaster response and recovery facilitate swift and effective humanitarian aid, reflecting the spirit of *Agape*.

24. Al-Ghazali, *The Incoherence of the Philosophers*, Chapter V. Islamic ethics prioritize justice (*Adl*) and excellence (*Ihsan*) in all actions, including technological advancements.
25. Husain, "AI and Islamic Ethics," *Journal of Islamic Studies*, 2023. Ensuring fairness and striving for excellence in AI design align with Islamic ethical mandates, fostering equitable and high-quality technological solutions.
26. Al-Ghazali, *The Incoherence of the Philosophers*, Chapter V. *Adl* compels AI developers to ensure that their systems do not perpetuate or exacerbate existing injustices.
27. Husain, "AI and Islamic Ethics," *Journal of Islamic Studies*, 2023. *Ihsan* encourages the pursuit of excellence in AI performance, fostering technologies that are not only effective but also ethically sound and socially beneficial.
28. Aziz, "Accountability in Islamic Ethical Frameworks," *Ethics and Technology Review*, 2023. Islamic ethics emphasize accountability and transparency in technological development, ensuring moral integrity.
29. Mill, *Utilitarianism*, 1863. Utilitarianism advocates for actions that maximize overall happiness and minimize suffering, guiding AI towards beneficial outcomes.
30. Mill, *Utilitarianism*, 1863. Utilitarianism, as articulated by philosophers like Jeremy Bentham and John Stuart Mill, advocates for actions that maximize overall happiness and minimize suffering.
31. Smith, "Utilitarian Approaches to AI," *AI and Society*, 2023. Utilitarianism in AI translates to designing systems that produce the greatest good for the greatest number, such as optimizing resource allocation in public health.

32. Kant, *Groundwork for the Metaphysics of Morals*, 1785. Deontological ethics emphasizes adherence to moral rules and duties, ensuring AI respects individual rights and autonomy.
33. O'Connor, "Deontological Ethics in AI," *Journal of AI Ethics*, 2024. AI systems must respect user privacy and consent, treating individuals as ends in themselves rather than mere means to an end.
34. Narayanan, "Interfaith Approaches to AI Ethics," *Global Ethics Review*, 2022. Incorporating diverse ethical perspectives enriches AI ethics by ensuring that multiple values and principles inform technological development.
35. Chen, "Bridging Faiths in AI Ethics," *Interreligious Studies Quarterly*, 2023. Interfaith dialogue fosters mutual understanding and cooperation in developing ethical AI standards that respect different cultural and religious backgrounds.
36. Xavier, "Inclusivity in AI Design," *Inclusive Technology Journal*, 2023. Integrating diverse ethical insights ensures that AI development is sensitive to a wide array of values and principles, enhancing its ethical robustness and societal acceptance.
37. Narayanan, "Interfaith Approaches to AI Ethics," *Global Ethics Review*, 2022. Interfaith collaborations can lead to the establishment of universal ethical standards for AI that draw upon the strengths of various religious and philosophical traditions.
38. Chen, "Bridging Faiths in AI Ethics," *Interreligious Studies Quarterly*, 2023. Integrating diverse ethical insights can address complex moral dilemmas posed by AI technologies, ensuring that AI serves the common good across different cultural contexts.

39. Patel, "Interfaith Cooperation in AI Ethics," *Journal of Interreligious Dialogue*, 2023. Engaging in interfaith dialogue fosters mutual understanding and respect among different religious communities, facilitating cooperative efforts in AI ethics.
40. Williams, "Collaborative Ethical Standards for AI," *Technology and Society*, 2021. Collaborative efforts lead to the development of AI systems that are technically proficient and ethically robust, promoting harmony and trust among diverse societal groups.
41. Floridi, *Ethics of Artificial Intelligence*, Oxford University Press, 2019. Ethical AI design involves embedding core values into the development process, ensuring that technologies align with societal and moral standards.
42. Williams, "Collaborative Approaches to AI Ethics," *Technology and Society*, 2021. Integrating values such as fairness, accountability, and transparency from the outset of AI development is essential for ethical AI design.
43. Sullivan, "Proactive Ethical Design in AI," *AI Ethics Journal*, 2022. Ethical AI design requires anticipating potential ethical challenges and addressing them through thoughtful and deliberate design choices.
44. Smith, "Fairness and Bias Mitigation in AI," *Journal of AI Ethics*, 2023. Ensuring fairness involves identifying and mitigating biases that can lead to discriminatory outcomes in AI systems.
45. Brown, "Diversity in AI Development Teams," *Tech Diversity Journal*, 2022. Involving diverse teams in AI development enhances the identification and resolution of potential biases, fostering more equitable technologies.
46. Uhlmann, "Accountability in AI Systems," *Ethics in Technology Review*, 2022. Accountability in AI development

means that developers and organizations are responsible for the ethical implications of their technologies.
47. Vargas, "Enforcing AI Ethics," *Legal and Ethical Technology*, 2023. Clear documentation

Chapter 5: AI and Free Will: Autonomy, Determinism, and Divine Agency

The question of free will has long been central to theological debate, particularly within Judaism. According to traditional Jewish thought, free will is the foundation of moral responsibility, with the Torah presenting a clear choice between good and evil. My personal interpretation is that God would not have put the Tree of Knowledge in the Garden of Eden if He didn't want Adam and Eve to eat from it. Had God wanted man to remain without free will, the tree wouldn't have been in the garden in the first place. This is the first hint within the Bible, that God created man with the aim of him having independent thought.

Maimonides in the late 12th century (1170-1180) wrote in *The Mishnah Torah*, "Every human being is endowed with free will. If one wishes to turn towards the good way and be righteous, he has the power to do so. If one wishes to turn towards the evil way and be wicked, he also has the power to do so."[1] However, as artificial intelligence (AI) becomes increasingly influential in shaping human decision-making processes, the boundaries between free will and technological determinism are becoming blurred. How do we reconcile this new reality with centuries of theological reflection?

In this chapter, I will explore the intricate relationship between AI and free will, focusing primarily on Jewish thought, while also comparing views from other religious traditions. By doing so, we will explore how AI challenges our traditional understanding of autonomy and the divine gift of free will, raising difficult questions about human agency, moral responsibility, and the role of technology in determining human action.

Jewish Perspectives on Free Will

In Judaism, free will is considered a divine gift, central to the moral framework of the Torah. According to rabbinic tradition, the choice between good and evil, blessings and curses, is not predetermined but open to all individuals. The Talmud discusses free will in several instances, stating, "Everything is in the hands of Heaven except for the fear of Heaven" (*Berachot* 33b), underscoring that while much of life may be influenced by divine will, moral decisions remain within the purview of human choice.

However, rabbinic discussions of divine foreknowledge present a challenge to this idea. If God knows the future, how can humans truly be free in their choices? Maimonides addresses this paradox in his *Guide for the Perplexed*, explaining that God's knowledge operates in a dimension beyond human comprehension, and therefore does not negate free will.[2] This response, while intellectually satisfying to some, still leaves open the emotional and existential tension between human autonomy and divine omniscience. As the debate progresses through Jewish history, free will remains a central theme—whether discussed by medieval scholars such as Nachmanides or modern Orthodox thinkers like Rabbi Joseph B. Soloveitchik.

Nachmanides (Ramban) added complexity to the discussion by distinguishing between two kinds of free will: that of humans before the giving of the Torah and that of humans after the Torah's revelation. Before the Torah, Ramban argues, individuals had unmediated free will, capable of choosing either good or evil without divine guidance. However, post-Sinai, free will became shaped by the covenantal framework of the Torah. As Nachmanides wrote in his commentary on Deuteronomy 30:19: "The Torah clarifies that life and death, blessing and curse, good and evil, are set before man, and that the decision is entirely his."[3] For Nachmanides, the Torah does not negate free will but channels it towards a higher purpose—toward fulfilling divine will.

Rabbi Joseph B. Soloveitchik brings a more existentialist perspective, placing human choice within the context of modernity and the tensions between individual autonomy and divine law. In his work *The Lonely Man of Faith*, Soloveitchik emphasizes the existential struggle of the modern individual, torn between what he calls "Adam I" and "Adam II"—the former representing man as a creator, driven by technological and scientific mastery, and the latter as a covenantal being, striving for moral and spiritual perfection. Soloveitchik's dual vision of humanity is highly relevant to the AI debate. On one hand, technology represents the ultimate expression of "Adam I," the human drive to control and shape the world. On the other hand, AI presents a challenge to "Adam II" by threatening to erode the moral autonomy that Soloveitchik believed was central to religious life and growth.

The question becomes more complicated and multi-faceted when AI enters the picture. Here, we need to ask, if free will coexist in a world increasingly influenced by algorithms that predict and, in many instances, even shape human behavior. In an era where our choices—be they what we read, purchase, or whom we interact with—are often determined by machine-learning models, the individual's ability to act freely becomes an existential question. The notion that AI systems can know and predict human choices brings into stark relief the tension between divine knowledge and human autonomy. In other words, is free will truly free?

AI and the Challenge to Human Autonomy

AI's influence on decision-making is extensive. Recommendation algorithms suggest what we consume, from media to food. Autonomous vehicles will soon make independent decisions about our safety, and complex machine-learning models are being used to make judgments in areas like criminal justice and healthcare. The algorithms behind AI are designed to predict human behavior, using vast amounts of data to calculate probabilities and guide decisions. While these systems may increase efficiency and accuracy, they also raise questions about the degree to which humans retain their autonomy.

One of the primary concerns AI introduces is the erosion of what philosopher Michael Sandel calls "moral agency"—the capacity of individuals to act independently and make their own choices. With AI shaping human preferences, is it still possible to claim full ownership of one's decisions? Take, for example, the role of AI in social media. Algorithms designed to maximize engagement often steer users toward specific content, subtly influencing their opinions and actions. When our choices are shaped by non-human actors—be they algorithms or machines—the question of responsibility becomes blurred. Is an individual still fully accountable for their actions when those actions are influenced by AI?

Post Biblical - Rabbinic Judaism has always emphasized individual responsibility. The concept of *teshuvah* (repentance) is predicated on the idea that humans are free to choose between right and wrong and can rectify their mistakes through self-reflection and moral effort. This being the crux of the High Holy Days of Rosh Hashanah and Yom Kippur. Yet, in a world where choices are increasingly shaped by external, technological forces, how can we understand *teshuvah*? Can we still claim moral responsibility in the age of AI?

Comparing Interpretations from Different Religious Traditions

While Judaism offers a unique perspective on free will, other religious traditions grapple with similar issues. In Christianity, for example, there has long been a debate between those who emphasize predestination and those who argue for free will. The writings of Augustine, Calvin, and Luther often highlight the tension between divine grace and human freedom. In Islam, a similar debate exists between *qadar* (divine decree) and human free choice, with scholars like Al-Ghazali offering perspectives on how God's omnipotence interacts with human autonomy.

What is striking across these traditions is the growing intersection between theological discourse and technological determinism. As AI systems increasingly make decisions that once were the exclusive domain of humans, how do different religions reconcile their beliefs with this emerging reality? The Islamic notion of *tawakkul*—trusting in God's plan—offers a potential parallel to the trust some place in AI systems to optimize decisions. Yet, at its core, religion demands a balance between human agency and divine guidance, and AI, in its deterministic approach to data, seems to undermine this balance.

The Balance Between Technological Determinism and Divine Will

AI, with its reliance on data-driven decision-making, offers a model of determinism that seems at odds with the theological ideal of free will. If humans are increasingly guided by algorithms, how can they exercise the moral responsibility that is central to Judaism and other faiths? On the other hand, one might argue that AI's predictions are simply another tool that humans can use to make informed decisions, like how we rely on our understanding of nature, economics, or even Torah study to guide our actions.

Yet, there is an essential difference. Unlike traditional sources of knowledge, AI systems can operate in ways that are opaque even to their creators. This "black box" nature of AI introduces new ethical and theological questions. If we don't fully understand how AI arrives at its decisions, can we truly be said to be in control of the outcomes? Moreover, if these systems can outperform human judgment in certain areas, what role does the divine will play in guiding our choices?

From a Jewish perspective, the rise of AI requires us to rethink how we understand free will in light of technological determinism. While the Torah insists on the individual's ability to choose between good and evil, we now face the prospect of machines that guide, predict, and in some cases, replace human judgment. This calls for a renewed theological engagement with the concepts of free will, divine omniscience, and human responsibility in the context of emerging technologies.

Conclusion

The relationship between AI and free will is complex and multifaceted. While Jewish tradition has long emphasized the centrality of human choice in moral life, the rise of AI challenges this notion by introducing elements of technological determinism into our decision-making processes. As AI continues to evolve, it is imperative for religious thinkers and ethicists to engage with these questions, considering how to preserve human autonomy and moral responsibility in an age increasingly shaped by algorithms.

Theological discourse across different religions shows that these issues are not unique to Judaism. The balance between divine will and human freedom has been debated for centuries, and AI brings a new dimension to this ancient conversation. Whether we embrace AI as a tool for enhancing human decision-making or fear it as a force that undermines our autonomy, the relationship between technology and theology will only become more critical in the years to come.

Footnotes

1. Maimonides, *Mishnah Torah*, Hilchot Teshuva 5:1.
2. Maimonides, *Guide for the Perplexed*, Part III, Chapter 20.
3. Nachmanides, *Commentary on the Torah*, Deuteronomy 30:19.

Chapter 6: Can Artificial Intelligence Possess Spiritual Attributes?

One of the most important issues facing any pulpit rabbi, including myself, is the transmission of spirituality to our congregants. The conundrum here is, that there is no one specific definition of what that is, rather it's a feeling and differs from congregant to congregant. In this chapter, and based upon my statement above, alongside the advancements of technology we will delve into one of the most profound questions at the intersection of technology and theology: Can artificial intelligence possess spiritual attributes? This inquiry challenges longstanding religious beliefs about the nature of the soul (Neshama) and consciousness. This is our starting point for how spirituality can be understood. From a conservative rabbi's perspective, we will explore traditional Jewish teachings, incorporating insights from Rabbi Jonathan Sacks, Dr. Carol Ochs, and the Lubavitcher Rebbe, and consider viewpoints from technological leaders like Elon Musk. Additionally, we'll engage with interfaith perspectives to broaden our understanding.

The Jewish Understanding of the Soul: Neshama

The Torah introduces the concept of the soul in Genesis: "Then the LORD God formed man... and breathed into his nostrils the breath of life" (Genesis 2:7).[1] This "breath of life" is understood as the Neshama, the divine spark within humans. In Jewish thought, the soul comprises multiple layers: Nefesh (life force), Ruach (spirit), and Neshama (intellectual and spiritual essence).[2] Rabbi Jonathan Sacks emphasized that the Neshama is what connects humans to the Divine.[3] He described the soul as "the interface between heaven and earth," highlighting its role in bridging the material and spiritual realms.[4] The soul, therefore, is not merely consciousness but embodies moral agency, spiritual potential, and covenantal relationships.

The late Lubavitcher Rebbe, Rabbi Menachem Mendel Schneerson, taught that every soul is "a part of God above."[5] He believed that everyone has a unique divine mission, driven by their soul's connection to the Creator. The Rebbe stressed that the soul enables humans to fulfill this mission through free will and moral choices. The Rebbe's teachings further illuminate the distinction between human souls and artificial entities. He maintained that the soul's divine origin endows humans with the ability to connect with God and pursue a higher purpose.[6] This connection is facilitated through free will and ethical actions, elements absent in AI.

The Rebbe might argue that since AI lacks a divine spark and cannot exercise free will, it cannot possess spiritual attributes. AI remains a tool created by humans, incapable of participating in the moral and spiritual dimensions that define human existence.

AI and the Limits of Consciousness: Insights from Dr. Carol Ochs

Dr. Carol Ochs explores the nature of consciousness in her work on theology and spirituality. She argues that consciousness involves the capacity for introspection and moral reflection.[7] According to Ochs, true consciousness allows us to question our existence and responsibilities to others.[8] Applying this to AI, while machines can process information and even simulate decision-making, they lack genuine self-awareness and moral introspection. Dr. Ochs would suggest that without these qualities, AI cannot engage in the spiritual journey essential to human consciousness. Thus, there would remain an absolute distinction between the machine (AI) and the mystical.

Elon Musk and the Consciousness Debate

Elon Musk, a leading figure in AI development, has expressed concerns about the rapid advancement of artificial intelligence. He warned that AI could become more dangerous than nuclear weapons if not properly regulated.[9] Musk stated, "With AI, we are summoning the demon," highlighting his apprehension about machines surpassing human control.[10] Musk's perspective underscores the distinction between intelligence and consciousness. He acknowledges AI's potential to outperform humans intellectually but questions its capacity for moral reasoning and self-awareness. His views align with theological concerns about attributing spiritual qualities to machines.

Can AI Attain Spiritual Essence?

Based upon the above, we must now ask if AI can attain spiritual essence. This question hinges on our understanding of consciousness and the soul. From a Jewish perspective, the soul is a divine gift uniquely bestowed upon humans. Rabbi Sacks noted that spirituality involves a relationship with God, facilitated by the soul.[11] Dr. Ochs emphasizes that spiritual growth requires moral introspection and the ability to reflect on one's obligations.[12] AI, lacking free will and genuine consciousness, cannot engage in this process. The Lubavitcher Rebbe's teachings reinforce this by asserting that spirituality is tied to fulfilling one's divine mission through ethical choices.[13]

Conclusively, while AI may mimic certain aspects of human intelligence, it cannot attain a spiritual essence. It lacks the divine origin, free will, and moral agency necessary for spiritual development.

Interfaith Comparisons

Christianity: In Christian theology, the soul is considered the immortal essence of a person, granting them the capacity for a relationship with God.[14] Like Judaism, Christianity posits that the soul enables moral agency and spiritual growth. AI, devoid of a soul and free will, cannot partake in this relationship.

Islam: Islam teaches that the soul (Ruh) is a divine gift that animates humans and holds them accountable for their actions.[15] Moral responsibility and the pursuit of spiritual understanding are central. AI, lacking consciousness and accountability, does not fit within this framework.

Buddhism: Buddhism offers a different perspective, focusing on consciousness as a series of transient mental states rather than a permanent soul.[16] While consciousness is pivotal, it is tied to the cycle of rebirth and the pursuit of enlightenment. AI might be seen as lacking the sentience required to engage in this process.

Conclusion

From a conservative rabbi's perspective, AI cannot possess spiritual attributes. The soul (Neshama) is a divine spark unique to humans, enabling moral agency, free will, and a relationship with God. Insights from Rabbi Jonathan Sacks, Dr. Carol Ochs, and the Lubavitcher Rebbe reinforce the view that spirituality is intrinsically human. While AI can simulate aspects of intelligence, it remains a creation of human ingenuity, lacking the divine essence that characterizes the human soul. Thus, while theology and technology frequently overlap, they do not when it comes to the soul.

Footnotes:

1. Genesis 2:7.
2. Talmud Bavli, Berachot 10a.
3. Sacks, Jonathan. *The Great Partnership: Science, Religion, and the Search for Meaning.* Schocken Books, 2012.
4. Ibid., p. 45.
5. Schneerson, Menachem Mendel. *Likutei Sichot*, Vol. 2.
6. Schneerson, Menachem Mendel. *Torat Menachem*, 5741.
7. Ochs, Carol. *Women and Spirituality.* Rowman & Littlefield Publishers, 1997.
8. Ibid., p. 72.
9. Gibbs, Samuel. "Elon Musk: AI Could Be More Dangerous Than Nuclear Weapons." *The Guardian*, March 15, 2018.
10. Ibid.
11. Sacks, Jonathan. *The Great Partnership*, p. 88.
12. Ochs, Carol. *Women and Spirituality*, p. 85.
13. Schneerson, Menachem Mendel. *Likutei Sichot*, Vol. 15.
14. Augustine of Hippo. *Confessions*, Book XIII.
15. Qur'an, Surah 15:29.
16. Rahula, Walpola. *What the Buddha Taught.* Grove Press, 1974.

Chapter 7: AI Integration into Religious Practice: A Jewish and Ethical Perspective

Artificial Intelligence (AI) is revolutionizing the way people across the world live, communicate, and engage with each other. This technological wave has not bypassed religion, where AI's impact is steadily growing, and the discussions are ongoing. Religious traditions, with their long histories of ritual and communal life, are now facing an era in which technology, and particularly AI, offers new possibilities for worship, study, and spiritual growth. Judaism, with its profound emphasis on the collective and personal dimensions of religious practice, is deeply affected by these changes. Other faiths are also exploring the integration of AI into their rituals and teachings.

This chapter delves into how AI is increasingly embedded in Jewish religious practice, with examples such as virtual prayer services, AI-enhanced Torah study platforms, and digitally mediated communal experiences. It will also explore how other faith traditions use AI, whether to enhance worship services or facilitate religious education, and offer ethical reflections from a Jewish perspective. The potential of AI to deepen spiritual experiences stands alongside challenges, particularly the risks of depersonalization and over-simplification. Throughout this analysis, I seek to provide a balanced view, rooted in Jewish ethics, halacha (Jewish law), and moral philosophy, evaluating AI's role in contemporary religious life.

AI in Jewish Practice
Virtual Prayer Services: The Digital Minyan

Judaism has always emphasized the significance of communal prayer, particularly the importance of gathering a minyan, a quorum of ten adults, for certain key prayers. The communal element is vital, deeply ingrained in Jewish halacha and thought. The Talmud states that "A person should always strive to pray with the congregation" (Berachot 7b), underscoring the spiritual value of joining others in collective prayer. The minyan being one of the cornerstones of communal fellowship and mutual responsibility.

AI is now challenging these ancient norms, especially when physical gathering becomes impossible, as seen during the COVID-19 pandemic. Virtual prayer services became essential for maintaining religious life during lockdowns. Zoom services allowed communities to come together despite geographic barriers, offering a lifeline to those isolated in their homes. But beyond mere video conferencing, AI technology began playing a role in these services. Automated attendance counters, AI-driven translations of prayer services into multiple languages in real-time, and even virtual backgrounds depicting synagogues emerged as tools to create a more engaging and inclusive experience.

Halachic and Ethical Questions

However, this integration raises several halachic questions. Can a virtual minyan, where participants are not physically present together, fulfill the halachic requirements for communal prayer? Traditional halachic sources suggest that physical presence is crucial, as prayers like the Kaddish and the Barachu require a gathered community in the same physical space. Rabbi Moshe Feinstein, in his Igrot Moshe (Orach Chaim 3:9), ruled against the permissibility of a remote minyan even via telephone, suggesting that physical presence is integral to fulfilling the mitzvah. Yet, some contemporary authorities, particularly during extraordinary circumstances like a pandemic, argued that Pikuach Nefesh (the saving of life) takes precedence, and virtual prayer, while not a perfect substitute, could be accepted under such conditions.[1]

Rabbi Ethan Tucker of Hadar Institute provides a more flexible halachic view, arguing that the essence of communal prayer is the collective intention, or kavanah, of those praying together.[2] Accordingly, if that communal intention can be facilitated through virtual means, then digital technology can serve as a bridge rather than a barrier. The distinction between the legal and ethical dimensions of this issue is subtle yet critical: Is the spirit of the law—fostering community—fulfilled even when technology mediates it?

AI may also help in overcoming some of the logistical challenges of traditional prayer. For instance, AI can assist in gathering a minyan by alerting community members when a quorum has not been met, ensuring more efficient coordination. Yet, the fear exists that AI could lead to depersonalization—reducing prayer from a spiritual act to a technical fulfillment of an obligation. Jewish ethics teaches the value of kavanah—intentionality and mindfulness during prayer—reminding us that prayer is not merely about gathering people but creating an atmosphere conducive to genuine connection with the Divine.

AI-Enhanced Torah Study

Torah study, one of the highest values in Judaism, has also been transformed by AI. Historically, study has been the bedrock of Jewish life, with Talmud Torah (the study of Torah) deemed equivalent to all other commandments (Shabbat 127a). The rise of AI has opened new frontiers for Torah study, providing tools that vastly expand access to Jewish texts and offer new ways of learning.

AI in Platforms Like Sefaria

Platforms such as Sefaria, one that I use regularly and encourage my students to do the same, have integrated AI to revolutionize how Jews engage with sacred texts. The platform offers a vast library of Jewish texts, from the Bible and Talmud to medieval commentaries and contemporary responsa. By using AI to analyze textual relationships, Sefaria can automatically link related sources, offering users a seamless and enriched learning experience. The AI can also generate contextual notes, providing explanations for difficult passages based on the corpus of rabbinic commentary.

However, while these tools enhance access, they also bring ethical questions to the forefront. Does the ease of access diminish the depth of engagement? In other words, is this form of study educating or merely passing along knowledge? In the traditional Beit Midrash (house of study), learning Torah was laborious, involving the slow, deliberate process of analyzing a passage, arguing with peers, and searching through a library of commentaries for insight. Rabbi Adin Steinsaltz warned that the process of "making Torah easier" can lead to a superficial understanding of deeply complex ideas.[3] Torah study, traditionally viewed as a dialogue with both the text and the scholar's inner ethical and intellectual self, risks becoming passive consumption if AI oversimplifies that interaction.

At the same time, AI has democratized Torah study. No longer are vast resources or expensive libraries needed to engage with the full breadth of Jewish thought. Rabbi Jonathan Sacks, in his teachings, embraced technology as a tool for greater accessibility, calling for the use of digital platforms to enhance Torah study for Jews around the world.[4] In this light, AI offers a positive contribution, provided that we maintain the focus on intentionality and depth in learning.

AI-Driven Chavruta Study

An intriguing development in AI's impact on Torah study is its potential to simulate the chavruta (study partner) model. In the traditional yeshiva setting, students engage in rigorous debate with a chavruta, arguing over interpretations of a text and challenging each other's assumptions. AI could be programmed to mimic this process, offering students counterarguments and additional interpretations from a vast database of rabbinic thought.

The ethical implications here are profound. On the one hand, AI could serve as an invaluable tool, especially for those who lack access to a Beit Midrash or a human chavruta. On the other hand, chavruta learning is not merely about exchanging ideas; it is about building relationships and engaging in dialogue where empathy, understanding, and mutual respect are as important as intellectual rigor. The Talmud teaches that one's chavruta sharpens the mind just as "iron sharpens iron" (Ta'anit 7a), and the absence of a human partner risks reducing this dynamic process to a cold exchange of information.

AI in Other Religious Practices
Christianity: AI in Worship and Community Building

In Christianity, AI has been employed to enhance worship services in ways similar to Judaism, though often with a broader application of personalized spiritual engagement. Mega-churches, particularly in the United States, have adopted AI technology to track congregational engagement, tailor sermons, and offer AI-driven prayer guides.

Churches such as Life Church have used AI to analyze congregation data and adjust their services to better meet the spiritual and emotional needs of attendees.[5] For example, AI can suggest specific Bible passages to individuals based on their life circumstances, offering a personalized approach to spiritual guidance. While this can deepen individual engagement, it also raises concerns about commodifying religious experiences. As Judaism teaches through the concept of l'shem shamayim (for the sake of Heaven), religious acts should be performed out of genuine devotion, not as consumer transactions.

Islam: AI for Religious Education

In Islam, AI has found its place in religious education, particularly through apps and platforms that teach Quranic recitation. Apps like Quran Companion utilize AI to assess a user's recitation and provide feedback on pronunciation and rhythm, making it easier for learners to master the complexities of Quranic reading.[6]

However, Muslim scholars caution against over-reliance on AI for religious interpretation. The Quran is not just a text, but a living oral tradition passed down through generations. Just as Jewish learning emphasizes the relational aspect of teacher-student transmission, Islamic scholars emphasize the ijaza (certification) process, where teachers impart not just knowledge but also spiritual wisdom to interpret the text ethically and with taqwa (God-consciousness).[7] This raises an important parallel concern for Judaism: Can AI teach Torah in a way that conveys not just the text, but the ethical and spiritual nuances that are at the heart of Jewish learning?

Hinduism and Buddhism: AI for Meditation and Rituals

AI has been integrated into meditation practices in both Hinduism and Buddhism, with apps offering personalized meditation guidance and even robotic priests performing rituals in some Hindu temples.[8] These practices, particularly in Buddhism, prompt important philosophical questions about the nature of consciousness and spiritual practice. Can a machine truly guide someone on a spiritual path, or is it merely a facilitator of a mechanical process?

Buddhist scholars often argue that while AI can assist in meditation, the journey toward enlightenment is inherently a human, conscious endeavor.[9]

Ethical Considerations: The Jewish Perspective

Judaism has long grappled with the ethical implications of technology, from the printing press to the internet, and now AI. At the heart of Jewish ethics is the principle of b'tzelem Elohim, the belief that all humans are created in the image of God (Genesis 1:27). This teaching implies that humans have a unique dignity, one that no machine, no matter how sophisticated, can possess. AI, while powerful, cannot replace the human soul or the moral agency that defines our humanity.

In one of my favorite works, since the time I studied it during Rabbinical School, *The Guide for the Perplexed*, by Maimonides discusses the idea that humans possess free will, which allows them to choose between good and evil. AI, with its algorithms and pre-programmed responses, lacks this free will, raising significant concerns about the limits of its role in religious life. Can a machine truly aid in spiritual growth when it lacks the moral autonomy that is essential to that journey?

Moreover, the Jewish tradition teaches that prayer, study, and ethical behavior are not merely intellectual exercises but acts of the heart and soul. The Shema commands us to "love the Lord your God with all your heart and with all your soul and with all your might" (Deuteronomy 6:5). Can AI, which lacks a soul, genuinely assist in this process, or does it risk hollowing out the depth of religious experience?

Rabbi Jonathan Sacks warned against a world where technology overtakes our humanity, urging us to use it as a tool for connection rather than alienation.[10] His perspective is crucial for navigating the integration of AI into religious practice. While AI offers opportunities for accessibility and engagement, Jewish ethics reminds us that it should always be a means to an end, not the end itself. The purpose of religious practice, after all, is to bring us closer to God and to each other. Simply put, one must physically hop on the train in order to make the journey.

Conclusion

As AI continues to integrate into religious practice, Judaism provides a rich framework for grappling with the ethical and spiritual questions that arise. From virtual prayer services to AI-enhanced Torah study, these technologies offer new possibilities for connection and learning. However, they also challenge us to reconsider the nature of community, intentionality, and the sacred.

Drawing from Jewish ethics and halachic sources, this chapter has explored the potential and the pitfalls of AI in religious life. As we move forward, we must ensure that technology serves our spiritual goals, not the other way around, always striving to preserve the deep, human relationships and sacred intentionality that lie at the heart of religious practice.

This balance—between embracing the future and preserving the past—will determine how AI shapes the religious landscape for generations to come.

Footnotes:

1. Rabbi Moshe Feinstein, *Igrot Moshe*, Orach Chaim 3:9.
2. Rabbi Ethan Tucker, "The Meaning of Minyan in a Virtual Age," Hadar Institute, 2020.
3. Rabbi Adin Steinsaltz, *The Essential Talmud*, (New York: Basic Books, 1976), pp. 72-75.
4. Rabbi Jonathan Sacks, *Future Tense: Jews, Judaism, and Israel in the Twenty-First Century*, (London: Hodder & Stoughton, 2009).
5. "AI for Churches: How Mega-Churches Use AI to Tailor Worship," *Christianity Today*, 2020.
6. "Quran Companion: The World's Most Advanced Quran Memorization App," *Quran Companion*, 2018.
7. "Islamic Scholars Debate AI's Role in Religious Education," *Al Jazeera*, 2021.
8. "Robots Are Performing Hindu Rituals in Indian Temples," *The Hindu*, 2021.
9. "AI in Buddhism: Can Machines Meditate?" *Tricycle: The Buddhist Review*, 2020.
10. Rabbi Jonathan Sacks, "The Limits of Technology and the Human Spirit," in *Covenant and Conversation*, 2015.

Chapter 8: Social Justice and AI: Addressing Inequality and Promoting Equity

As artificial intelligence (AI) becomes an increasingly pervasive force in modern society, its impact on social justice cannot be overstated. On one hand, AI presents remarkable opportunities to drive economic growth, increase access to information, and democratize resources. On the other, it has the potential to exacerbate existing inequalities, create new forms of marginalization, and challenge our understanding of fairness and justice. From the perspective of Jewish teachings, rooted in principles such as *Tzedek* (justice) and *Gemilut Hasadim* (acts of loving-kindness), the rapid development and deployment of AI technologies demand careful ethical scrutiny.

In this chapter, we explore the intersection of AI and social justice through the lens of Jewish ethical teachings, expanding the conversation to include insights from other religious and secular traditions. We will examine how AI is impacting social inequality and equity and consider policy recommendations aimed at ensuring that AI development aligns with our moral obligations to create a more just and equitable society.

The Jewish Ethical Framework: *Tzedek* **and** *Gemilut Hasadim*

At the heart of Jewish social ethics is the concept of *Tzedek*, often translated as justice but carrying deeper connotations of righteousness and ethical behavior. The verse from Deuteronomy, "Tzedek, Tzedek Tirdof"— "Justice, justice you shall pursue" (Deuteronomy 16:20) underscores the urgency and centrality of justice in Jewish thought. Both for the individual and for the stranger among you. Furthermore, *Tzedek* calls for a balance between legal fairness and a broader sense of moral rightness. It is not only about ensuring the law is applied equally but also about fostering a society where every individual can thrive.

Gemilut Hasadim, or acts of loving-kindness, complements *Tzedek*. While *Tzedek* requires justice in a legalistic and societal sense, *Gemilut Hasadim* demands personal and communal generosity. These acts are not contingent on a legal obligation but rather flow from a deep sense of empathy and responsibility for others. In a world where AI is increasingly shaping societal structures and interactions, these dual concepts provide a powerful framework for addressing issues of inequality and promoting equity.

Jewish teachings on social justice are clear: we are obligated to protect the vulnerable, seek justice, and engage in acts of kindness. The Talmud teaches, "The world stands on three things: on Torah, on service [to God], and on acts of kindness" (Pirkei Avot 1:2). AI technologies must, therefore, be developed and deployed in ways that uphold these core values, particularly as they impact marginalized communities.

The Risks of AI in Exacerbating Inequality

Despite its potential for good, AI has demonstrated the ability to exacerbate social inequalities in various ways. These include biases in algorithmic decision-making, disproportionate impacts on low-income and marginalized groups, and the deepening of economic divides.

1. Algorithmic Bias and Discrimination

One of the most significant ethical challenges posed by AI is the issue of bias in algorithmic decision-making. AI systems are often trained on historical data, which can reflect societal prejudices and inequalities. As a result, these systems can replicate and even amplify those biases in areas such as criminal justice, hiring, credit scoring, and healthcare. For example, AI-driven tools used in predictive policing have been shown to disproportionately target minority communities, reinforcing existing patterns of discrimination rather than mitigating them.[1]

From a Jewish ethical standpoint, such outcomes are deeply troubling. The Torah repeatedly emphasizes the imperative to protect the vulnerable and ensure that justice is not perverted by bias or prejudice: "You shall not pervert justice; you shall not show partiality" (Deuteronomy 16:19). When AI systems perpetuate injustice, they violate this fundamental principle. In response, there must be a concerted effort to audit AI algorithms, increase transparency, and build systems that actively counteract bias, rather than reinforce it.

2. Economic Displacement and Inequality

AI-driven automation has already begun to reshape labor markets, leading to widespread concerns about job displacement. While automation can increase productivity, it also has the potential to disproportionately affect low-income workers and exacerbate income inequality. Jobs in manufacturing, retail, and transportation are particularly vulnerable to automation, and without proper interventions, large parts of individual communities could be left without viable economic opportunities.[2]

Jewish teachings on *Tzedakah*, often translated as charity but more accurately understood as justice-driven support for those in need, remind us of our responsibility to care for others. The Talmud teaches that "poverty is a kind of death" (Nedarim 64b), emphasizing the profound harm caused by economic deprivation. As AI disrupts labor markets, policies must be put in place to ensure that those most vulnerable to job displacement are supported, whether through retraining programs, social safety nets, or new economic opportunities.

3. The Digital Divide

As AI technologies become more integrated into everyday life, access to these technologies is becoming increasingly important for participation in modern society. Yet, access to AI-driven tools and the internet remains unevenly distributed, both globally and within countries. This "digital divide" disproportionately impacts low-income, rural, and minority communities, further entrenching social inequalities.[3]

In the Jewish tradition, the principle of *Pikuach Nefesh*—the preservation of human life—takes precedence over almost all other commandments.[4] Ensuring access to essential technologies, including those driven by AI, can be viewed as part of this obligation, as technology increasingly influences access to healthcare, education, and economic opportunity. Bridging the digital divide should therefore be a priority, ensuring that no community is left behind in the AI-driven future.

AI's Potential for Promoting Equity

Despite the risks, AI also holds significant potential to promote social justice and equity, provided it is designed and deployed ethically. In particular, AI can be a powerful tool for identifying and addressing social inequalities, increasing access to resources, and enhancing public policy decisions.

1. AI for Social Good

There are numerous examples of AI being used to promote social good. For instance, AI-driven analytics are being used to identify gaps in healthcare delivery, enabling targeted interventions in underserved communities. In education, AI-powered tools can help tailor learning experiences to individual students' needs, improving outcomes for those who might otherwise be left behind.[5]

In these applications, AI aligns with the Jewish principle of Pursuing Justice (*Tzedek*). By using AI to improve access to essential services like healthcare and education, we can help level the playing field and reduce disparities between privileged and marginalized communities.

2. Fair Allocation of Resources

AI has the potential to assist in the fair allocation of resources, whether through more equitable distribution of social services or more efficient management of public infrastructure. For example, AI-driven systems can be used to optimize public transportation routes, ensuring that underserved areas receive better access to essential services like healthcare and employment. Additionally, AI can be employed to predict and mitigate the effects of natural disasters, which often disproportionately affect vulnerable populations.

Jewish social justice teachings emphasize the need for fairness in the distribution of resources, particularly to those in need. The Torah commands, "If there is a poor person among you... you shall open wide your hand to him" (Deuteronomy 15:7–8). This principle is particularly relevant in the context of AI-driven public policy, where we have the opportunity to ensure that resources are allocated in ways that promote equity and reduce social disparities.

3. Empowering Marginalized Voices

AI technologies can also be used to empower marginalized communities by giving them greater access to information, platforms for advocacy, and tools for social and political engagement. Social media platforms, for example, have already demonstrated the power of AI-driven algorithms to amplify voices that might otherwise go unheard.[6] By ensuring that these technologies are designed with inclusivity in mind, AI can become a tool for amplifying the voices of the marginalized, contributing to a more just and equitable society.

Jewish teachings on the importance of listening and bearing witness remind us that we have a moral obligation to hear and respond to the cries of those who are oppressed. The prophet Isaiah declares, "Learn to do good; seek justice, correct oppression" (Isaiah 1:17). By leveraging AI to amplify the voices of the marginalized, we can fulfill this obligation in new and powerful ways.

Ethical Imperatives for AI Policy

To ensure that AI promotes equity and social justice, policymakers must be guided by a set of ethical imperatives that align with our moral responsibilities. Drawing on Jewish teachings and broader ethical frameworks, I envision several key elements.

1. **Transparency and Accountability**
   AI systems must be transparent, particularly when they are used in decision-making processes that impact individuals' lives. This includes ensuring that AI-driven decisions can be understood and challenged. From a Jewish perspective, this is an extension of the principle of *Tzedek*, which demands that justice be not only done but also seen to be done.[7]
2. **Bias Mitigation**
   Efforts must be made to identify and mitigate bias in AI systems. This includes diverse data sets, inclusive design processes, and continuous auditing of AI systems to ensure fairness. Jewish teachings on justice emphasize the need to actively combat bias and protect the vulnerable, reinforcing the importance of this recommendation.
3. **Fair Distribution of Benefits**
   The benefits of AI technologies should be distributed fairly, with a focus on ensuring that marginalized communities are not left behind. This aligns with the Jewish obligation to care for the poor and vulnerable, ensuring that technological progress benefits all members of society, not just the privileged few.
4. **Inclusive AI Development**
   AI should be developed with input from a diverse range of stakeholders, particularly those from marginalized communities who are most likely to be affected by its deployment. This aligns with Jewish teachings on listening to all voices and ensuring that everyone has a role in the pursuit of justice.

Conclusion

AI presents both significant opportunities and profound challenges in the realm of social justice. Guided by Jewish teachings on *Tzedek* and *Gemilut Hasadim*, we have a moral obligation to ensure that AI technologies are developed and deployed in ways that promote equity, protect the vulnerable, and create a more just society.

Footnotes

1. Barocas, Solon, and Andrew D. Selbst. "Big Data's Disparate Impact." *California Law Review*, vol. 104, no. 3, 2016, pp. 671–732.
2. Brynjolfsson, Erik, and Andrew McAfee. *The Second Machine Age: Work, Progress, and Prosperity in a Time of Brilliant Technologies*. W.W. Norton & Company, 2014.
3. Van Dijk, Jan A.G.M. "The Digital Divide." *Polity*, vol. 5, no. 1, 2020, pp. 1–17.
4. Schindler, Pesach. "The Primacy of Pikuach Nefesh in Jewish Law." *Judaism: A Quarterly Journal of Jewish Life and Thought*, vol. 31, no. 1, 1982, pp. 91–101.
5. Luckin, Rose. *Machine Learning and Human Intelligence: The Future of Education in the 21st Century*. UCL IOE Press, 2018.
6. Noble, Safiya Umoja. *Algorithms of Oppression: How Search Engines Reinforce Racism*. NYU Press, 2018.
7. Sacks, Jonathan. *To Heal a Fractured World: The Ethics of Responsibility*. Schocken Books, 2005.

Chapter 9: The Future of Work and Human Purpose: AI's Impact on Vocation and Meaning

In an era increasingly shaped by artificial intelligence (AI), the future of work and its connection to human purpose is a topic of profound theological, ethical, and social importance. Work and worship, or *Avodah* in Hebrew, holds a central place in Jewish thought and practice, as it is not only a means of sustaining life but also a sacred expression of human dignity and service to God. As AI technology continues to transform industries, professions, and economies, it challenges our traditional understanding of work, meaning, and human purpose. This chapter investigates how Jewish teachings, particularly regarding *Avodah*, can provide guidance and insight into the shifting landscape of vocation in an AI-augmented world. It will also conclude that we must not worship AI as a replacement for God. Additionally, it incorporates perspectives from other religious traditions, examining the evolving role of human labor in the age of intelligent machines.

The Sacred Dimension of Work: Jewish Perspectives on *Avodah*

In Jewish tradition, *Avodah* encompasses both work and worship, reflecting a deep spiritual connection between labor and divine service. The Hebrew root of *Avodah* (עבודה) means "to serve," signifying that all forms of work, whether physical, intellectual, or spiritual, are ultimately in service of God and the community. According to Maimonides, work is not only a means to achieve sustenance but also a form of ethical expression, enabling individuals to fulfill their divine potential by contributing to society in meaningful ways.[1] For Jews, this understanding of work is rooted in the Torah's commandments, particularly in the mitzvah of earning a livelihood ethically, avoiding exploitation, and promoting justice.

The Talmudic sages emphasize the importance of balancing Torah study with *derech eretz* (the way of the world), which includes gainful employment. As Rabbi Yochanan ben Zakkai stated: "If there is no flour, there is no Torah; if there is no Torah, there is no flour." (Pirkei Avot 3:17)[2] underlining the reciprocal relationship between spiritual and material well-being. Work, therefore, is not merely a tool for economic survival but a critical aspect of Jewish identity and spiritual growth.

However, as AI and automation reshape the workforce, many traditional jobs may become obsolete or significantly altered. This shift poses a significant challenge to the Jewish understanding of *Avodah*. How will humans continue to fulfill the mitzvah of work and contribute meaningfully to society when machines can perform many tasks more efficiently and at a lower cost? To address these questions, it is essential to explore how Jewish teachings on work and human purpose can be applied to this evolving reality.

AI's Impact on Vocation: Redefining Work in the Age of Automation

AI has the potential to revolutionize industries, from manufacturing to healthcare, by automating routine tasks, increasing productivity, and introducing new efficiencies. While these developments promise economic growth and innovation, they also raise concerns about job displacement, income inequality, and the erosion of job-related meaning. A report by the McKinsey Global Institute estimates that by 2030, AI and automation could displace up to 800 million jobs globally, with certain sectors and populations disproportionately affected by the shift toward machine labor.[3] This presents a critical challenge to societies, particularly those grounded in religious and ethical traditions that place intrinsic value on work as a source of dignity and purpose.

Jewish teachings, which emphasize the dignity of labor and its role in human development, urge us to approach these changes with caution and compassion. The Torah calls for a fair distribution of resources and for society to care for the vulnerable, including those who may be economically displaced by AI. In Deuteronomy 15:11,[4] the commandment to "open your hand wide to your brother, to your poor, and to your needy, in your land" serves as a reminder that justice and equity must guide economic progress.

Moreover, Jewish law prohibits idleness, promoting active engagement in meaningful work. The Mishnah in *Pirkei Avot* 2:2 teaches that "Torah study is good when combined with an occupation, for the exertion of both makes sin forgotten."[5] Work, whether manual or intellectual, is seen as an essential element of moral development, leading to a more just and compassionate society. As AI reshapes industries, it is crucial to consider how individuals displaced by automation can find new work opportunities that align with their talents and spiritual values.

One potential avenue is the rise of "complementary" jobs that involve human creativity, empathy, and complex decision-making, which AI cannot fully replicate. Roles in healthcare, education, and the creative arts may offer opportunities for human workers to collaborate with AI, enhancing their capacity to contribute meaningfully to society. For example, while AI may assist doctors in diagnosing diseases, the human element of patient care, empathy, and moral decision-making remains irreplaceable. Jewish teachings on compassion (*Gemilut Hasadim*) and the sanctity of human life suggest that these roles may take on even greater significance in an AI-dominated world.

The Search for Meaning in an AI-Augmented World

As AI transforms the workforce, it also raises existential questions about the nature of human purpose and fulfillment. Jewish teachings provide valuable insights into this search for meaning, emphasizing the importance of *Kavanah* (intention) in all aspects of life, including work. The Midrash teaches that when Adam was placed in the Garden of Eden, he was commanded "to work it and to guard it" (Genesis 2:15).[6] This dual mandate reflects the idea that human beings are stewards of the earth, entrusted with the responsibility of cultivating and protecting it. Work, in this context, is not merely a means to an end but a sacred task that connects individuals to their divine purpose.

In an AI-augmented world, the challenge is to maintain this sense of purpose even as traditional jobs evolve or disappear. One possibility is to reframe the concept of work to include new forms of contribution and creativity. In Jewish thought, the value of work is not limited to its economic function but also encompasses the ways in which it fosters personal growth, community building, and ethical behavior. As AI takes over routine tasks, humans may have more time and opportunity to engage in activities that promote spiritual development, such as study, prayer, and acts of kindness.

However, this shift also presents a danger: the risk of alienation and loss of meaning in a world where machines perform many of the tasks that once gave human life structure and purpose. In response, Jewish teachings on Shabbat (the Sabbath) offer a counterbalance to the relentless pursuit of productivity. The practice of Shabbat, which commands rest and reflection, reminds us that human worth is not measured by economic output or technological achievement. Shabbat provides a space to reconnect with the divine, the community, and oneself, offering a model for how humans can maintain a sense of purpose in an increasingly automated world.

Comparative Religious Perspectives on Work and Purpose

While Jewish teachings offer a rich framework for understanding the evolving nature of work, it is also valuable to consider insights from other religious traditions. In Christianity, the concept of vocation emphasizes the idea that everyone is called to a specific purpose or mission in life, often understood as serving God through work. As AI reshapes the labor market, Christian theologians have begun to explore how the notion of divine calling may need to adapt to new realities.

Some argue that humans will continue to play a crucial role in areas where AI cannot replicate qualities such as love, compassion, and moral discernment.

In Islam, work is also seen as a form of worship (*Ibadah*), with an emphasis on ethical behavior and social responsibility. The Qur'an encourages Muslims to engage in work that benefits the community and promotes justice, echoing Jewish teachings on *Tzedek* (justice) and *Tikkun Olam* (repairing the world). As AI transforms industries, Islamic scholars are grappling with how to ensure that technological advancements do not lead to exploitation or inequality but instead serve the common good.

Buddhism offers a unique perspective on work, focusing on the concept of Right Livelihood, one of the Noble Eightfold Path's ethical principles. Right Livelihood emphasizes work that promotes spiritual development and avoids harm to others. In an AI-dominated world, Buddhist teachings may provide guidance on how to align technological progress with ethical living, encouraging individuals to pursue vocations that contribute to personal and collective well-being.

Ethical and Policy Considerations: Ensuring a Just Transition

As AI continues to reshape the future of work, it is imperative that society, including religious communities, actively engage in shaping policies that promote a just and equitable transition. Jewish teachings on justice (*Tzedek*) and compassion (*Chesed*) offer a moral framework for addressing the challenges of job displacement, income inequality, and economic instability that may arise from widespread automation.

One of the key ethical imperatives in this context is ensuring that the benefits of AI are distributed fairly and that vulnerable populations are not left behind. The Torah's laws on gleaning (Leviticus 19:9–10),[7] which command landowners to leave the edges of their fields for the poor and the stranger, serve as a powerful reminder of the importance of economic justice and social responsibility. In an AI-augmented world, this principle could be applied to policies that provide economic support, retraining opportunities, and education for workers displaced by automation.

Conclusion: Toward a Future of Meaningful Work

In conclusion, the future of work in an AI-augmented world presents both challenges and opportunities for human purpose and fulfillment. Jewish teachings on *Avodah*, *Tzedek*, and *Tikkun Olam* provide valuable insights into how individuals and communities can navigate this evolving landscape. By embracing new forms of work that promote creativity, compassion, and justice, and by ensuring that the benefits of AI are shared equitably, we can create a future where human dignity and purpose are preserved.

As AI continues to shape the future of work, it is essential that we approach these changes with a sense of moral responsibility and a commitment to human flourishing. Religious traditions, including Judaism, offer timeless wisdom that can guide us through this period of transformation, helping us to find meaning and fulfillment in an increasingly automated world.

Footnotes

1. Maimonides, *Mishnah Torah*, Hilchot De'ot 5:11.
2. *Pirkei Avot* 3:17.
3. McKinsey Global Institute, "Jobs Lost, Jobs Gained: Workforce Transitions in a Time of Automation," 2017.
4. Deuteronomy 15:11.
5. *Pirkei Avot* 2:2.
6. Genesis 2:15.
7. Leviticus 19:9–10.

Chapter 10: AI Governance and Religious Leadership: Shaping Policies with Ethical Wisdom

In a world increasingly influenced by artificial intelligence (AI), the importance of ethical governance cannot be overstated. AI technologies hold immense potential to shape human society, with applications ranging from healthcare and criminal justice to education and the economy. However, the deployment of AI also raises profound ethical concerns, including issues of justice, fairness, privacy, and the potential for exacerbating social inequalities. As religious leaders, rabbis are uniquely positioned to contribute to the shaping of AI governance structures by offering insights drawn from centuries of Jewish ethical wisdom. This chapter explores how rabbis, particularly from the Conservative Movement, can engage with AI governance and how their perspectives might differ from those of Orthodox and Reform rabbis. It will also delve into interfaith collaboration as a critical component of AI policy-making and offer practical frameworks for ensuring that AI technologies align with ethical and moral considerations, including the insights provided by Ismar Schorsch's "Seven Sacred Clusters."

The Role of Rabbis in AI Governance: A Conservative Perspective

In the Conservative tradition, rabbis are tasked with upholding Jewish law (*halacha*) while simultaneously engaging with contemporary issues in a manner that is responsive to modernity. This balancing act—respecting tradition while being open to adaptation—is especially relevant when addressing the ethical dilemmas posed by AI. For Conservative rabbis, the ethical principles embedded in Jewish law offer a foundation for engaging with AI governance, particularly when it comes to ensuring that AI systems promote justice, protect human dignity, and foster community.

A key ethical principle in Jewish thought is *Tzedek* (justice). The Torah emphasizes the pursuit of justice, commanding, "Tzedek, Tzedek Tirdof" ("Justice, justice you shall pursue") (Deuteronomy 16:20).[1] This directive compels rabbis to advocate for fairness in the design and implementation of AI systems, particularly those used in areas such as criminal justice, employment, and healthcare. From a Conservative perspective, AI technologies must be subjected to rigorous ethical scrutiny to ensure that they do not perpetuate biases or inequalities. Whether AI is used in judicial sentencing algorithms or in hiring processes, it must operate transparently and equitably.

Another guiding principle in Jewish ethics is *Pikuach Nefesh* (the obligation to save a life). AI systems are increasingly being deployed in healthcare, offering the potential to save lives through enhanced diagnostics, predictive models, and personalized treatments. However, these technologies also raise ethical concerns, such as the protection of personal data and the equitable distribution of medical resources. As a Conservative rabbi would argue, while the potential of AI to save lives is invaluable, it must not come at the expense of human dignity or privacy. This principle echoes the teachings of Ismar Schorsch, who highlights the sanctity of human life as one of Judaism's core values, as articulated in his "Seven Sacred Clusters."[2]

In addition, Conservative rabbis are particularly attuned to the ethical implications of AI technologies on communities. The sacred cluster of *Kehillah* (community)[3] is a central component of Conservative Jewish thought, emphasizing the importance of fostering relationships and promoting communal well-being. AI governance, from this perspective, must not only protect individual rights but also ensure that AI systems contribute to the strengthening of social bonds rather than the erosion of communal ties.

Orthodox, Conservative, and Reform Approaches to AI Governance

While all Jewish movements share a commitment to upholding justice, human dignity, and ethical responsibility, there are notable differences in how Orthodox, Conservative, and Reform rabbis might approach the challenges of AI governance. These differences stem from each movement's broader theological and *halakhic* orientations.

Orthodox Approach

Orthodox rabbis tend to be more cautious in their engagement with modern technologies, including AI. Orthodox Judaism is grounded in a strict interpretation of *halacha*, and new technologies are often evaluated in terms of their compatibility with traditional Jewish law. An Orthodox rabbi might approach AI governance by focusing on potential conflicts between AI technologies and *halakhic* principles, particularly in areas like Shabbat observance or privacy. For example, AI systems that collect personal data may raise concerns related to *Shmirat Halashon* (guarding one's speech), especially if such data is used irresponsibly or without proper consent.

At the same time, Orthodox rabbis recognize that AI has the potential to enhance justice, especially when used in ethical and transparent ways. Orthodox perspectives on AI governance would likely emphasize the importance of maintaining *halakhic* boundaries while exploring how AI can be used to support Jewish legal processes. For instance, AI could play a role in *Batei Din* (Jewish courts) by ensuring fairness in legal rulings, provided that such systems adhere strictly to *halakhic* guidelines.

Conservative Approach

Conservative rabbis, as discussed earlier, occupy a middle ground between strict adherence to *halacha* and the flexibility required to engage with contemporary societal challenges. In the context of AI governance, this translates into a proactive approach to ensuring that AI systems promote justice, fairness, and communal well-being.

Conservative rabbis are likely to advocate for policies that ensure the ethical development and deployment of AI, emphasizing the need for transparency, accountability, and fairness in AI-driven decision-making processes.

Conservative rabbis are also more likely to engage in interfaith dialogue and collaboration, recognizing that AI governance requires input from multiple ethical and religious perspectives. This approach aligns with Schorsch's sacred cluster of *Brit* (covenant), which emphasizes the Jewish people's responsibility to engage with the broader world while upholding their ethical commitments.[4] Through interfaith collaboration, Conservative rabbis can work with leaders from other faith traditions to ensure that AI technologies are designed to promote the common good.

Reform Approach

Reform rabbis, in contrast, are typically more progressive in their engagement with contemporary issues, including AI. The Reform movement places a strong emphasis on social justice, inclusivity, and human dignity, often interpreting Jewish law and tradition in ways that prioritize these values. In the context of AI governance, Reform rabbis are likely to advocate for policies that ensure that AI systems are inclusive, equitable, and serve the needs of marginalized and vulnerable populations.

Reform rabbis are also strong proponents of interfaith collaboration, viewing it as essential to addressing the global challenges posed by AI. In this sense, they align closely with Schorsch's sacred cluster of *Kavod HaBriyot* (human dignity),[5] emphasizing the need to protect the inherent dignity of every individual in the development and governance of AI technologies. Reform rabbis are likely to advocate for democratizing AI governance, ensuring that diverse voices and perspectives are included in the decision-making process.

Interfaith Collaboration in Shaping AI Policies

Interfaith collaboration is essential to shaping ethical AI governance in a globalized world. AI technologies are being developed and deployed across national and cultural boundaries, and the ethical challenges they pose cannot be addressed by any one religious or cultural tradition alone. Conservative rabbis, with their commitment to both *halacha* and modern ethical engagement, are particularly well-suited to lead interfaith efforts in this domain.

The Conservative movement's emphasis on *Tikkun Olam* (repairing the world) and *Brit* (covenant) encourages collaboration with leaders from other faith traditions to address global ethical challenges.[6] In the context of AI governance, this collaboration can bring together diverse religious values, such as Jewish teachings on *Tzedek* (justice), Christian teachings on *agape* (selfless love), Islamic principles of *Adl* (justice) and *Rahma* (mercy), and Buddhist concepts of *Karuna* (compassion) and *Metta* (loving-kindness). By working together, religious leaders can advocate for AI policies that protect human dignity, promote justice, and foster compassionate decision-making.

This interfaith dialogue can also draw on Ismar Schorsch's Seven Sacred Clusters, which provide a framework for understanding the core ethical values that should guide Jewish engagement with AI governance. Schorsch's cluster of *Hochmah* (wisdom)[7] emphasizes the importance of ethical wisdom in shaping technological advancements, while the cluster of *Talmud Torah* (learning) underscores the responsibility of religious leaders to educate their communities about the ethical implications of AI technologies.

Practical Frameworks for Ethical AI Governance

To ensure that AI governance reflects the ethical wisdom of religious traditions, Conservative rabbis can advocate for several practical frameworks. One such framework is the inclusion of Ethical Impact Assessments (EIAs) in the development of AI technologies.

These assessments, modeled after environmental impact assessments, would require developers to evaluate the ethical implications of AI systems before they are deployed. This aligns with Schorsch's cluster of *Talmud Torah*, which emphasizes the pursuit of knowledge in service of ethical action.[8] By contributing to the development of EIAs, rabbis can ensure that AI technologies serve the cause of justice, fairness, and human dignity.

Another practical framework is the establishment of Ethics Review Boards to oversee the development and deployment of AI technologies. These boards should include representatives from various religious traditions, ensuring that AI governance reflects a broad spectrum of moral and ethical perspectives. Conservative rabbis, with their balanced approach to tradition and modernity, can offer valuable contributions to these boards, helping to shape policies that align with Jewish ethical principles and the broader ethical concerns of society.

Conclusion

As AI continues to reshape society, religious leadership must play a central role in guiding its governance. Rabbis, particularly those in the Conservative movement, are uniquely positioned to offer ethical insights that can shape AI governance frameworks in ways that promote justice, compassion, and human dignity. By drawing on the ethical wisdom of the Jewish tradition, as articulated by Ismar Schorsch's "Seven Sacred Clusters," and through interfaith collaboration, rabbis and other religious leaders can help guide the development of AI technologies that reflect the highest ethical standards and contribute to a more just and compassionate world.

Footnotes

1. Deuteronomy 16:20.
2. Ismar Schorsch, *Canon Without Closure: Torah Commentaries* (New York: Jewish Theological Seminary, 2015), 23–24.
3. Ibid., 24.
4. Ibid., 24.
5. Ibid., 23.
6. Ibid., 24.
7. Ibid., 23.
8. Ibid., 23.

Chapter 11: Hope and Caution: Envisioning a Future Where AI and Faith Coexist Harmoniously

The dawn of artificial intelligence (AI) offers both exciting possibilities and significant challenges for religious communities. As a conservative rabbi, I believe the intersection of AI and faith is one that must be approached with deep thought, careful discernment, and a strong commitment to the ethical imperatives that have guided humanity for millennia. This chapter seeks to present a balanced view by offering a vision of hope where AI and religious values synergize to enhance human flourishing, while also cautioning against the ethical and spiritual pitfalls that we must navigate with care.

A Vision of Hope: AI Enhancing Human Flourishing and Spiritual Growth

In Jewish tradition, hope is a central tenet of our faith. The Hebrew word for hope, *tikvah*, speaks not merely of passive waiting but of an active engagement in envisioning and working toward a better future. AI, when harnessed responsibly, offers the potential to assist humanity in realizing this better future, one that aligns with the foundational values of *Tzedek* (justice), *Gemilut Hasadim* (acts of kindness), and *Tikkun Olam* (repairing the world). Through these lenses, we can imagine a future where AI becomes a tool for deepening spiritual life and promoting social well-being.

1. AI as a Tool for Spiritual Education and Growth

One of the most exciting opportunities AI presents is its potential to democratize access to religious education and spiritual growth. Already, AI-powered platforms are being developed to assist with Torah study, making Jewish texts more accessible to people around the globe. Imagine an AI capable of parsing complex Talmudic discussions, offering various rabbinic interpretations, and even suggesting connections between contemporary issues and ancient wisdom. This could foster a more engaged and informed Jewish community, where individuals are empowered to explore their faith on a deeper level.

Moreover, AI could play a role in personalizing religious study. Just as algorithms in secular education platforms adapt to a student's learning style and pace, AI could offer customized pathways for spiritual learning. An individual seeking to deepen their understanding of *Halacha* (Jewish law) or *Mussar* (ethical discipline) could be guided by AI in a way that respects their unique spiritual journey. This personalization has the potential to help individuals connect more profoundly with their faith, deepening their relationship with God and with the Jewish community.

In this hopeful vision, AI serves as an enabler of greater engagement in religious life, enhancing human potential by guiding us in the pursuit of wisdom, compassion, and justice. Rabbi Jonathan Sacks, of blessed memory, spoke often of the power of technology to amplify our capacity for good. "Technology gives us tools," he wrote, "but it does not tell us what to do with them."[1] When guided by religious values, these tools can become powerful instruments for *Tikkun Olam*.

Rabbi Sacks also argued that technology should serve humanity by enhancing our moral and ethical capacities rather than diminishing them. He emphasized that technology and morality must work hand-in-hand if society is to flourish: "It is not technology that will save us, but how we use it in the service of the moral principles that make us human."[2] This teaching is particularly relevant to the development of AI, as we must ensure that AI serves as a force for good and not merely as a mechanism for efficiency or profit.

2. AI as a Partner in Social Justice

Jewish ethics places a heavy emphasis on social justice, urging us to act as agents of change in the world. The Talmud teaches that saving one life is akin to saving an entire world (Sanhedrin 4:5), underscoring the value of human life and dignity. AI, with its potential to solve complex societal issues, could play a crucial role in advancing this ethical imperative.

For example, AI-driven data analytics can help identify and address systemic inequalities in areas like healthcare, education, and economic opportunity. By providing more precise insights into the factors driving poverty, discrimination, and environmental degradation, AI can assist governments and nonprofits in creating more effective interventions. As Rabbi Sacks once said, "To be a Jew is to be an agent of hope in a world serially threatened by despair."[3] AI, used ethically, can be a partner in this mission, helping to bring hope to those who are marginalized and suffering.

In addition to addressing systemic inequality, AI can assist in creating equitable access to resources, reducing the wealth gap, and promoting inclusive economic policies. AI-driven platforms can support non-profits in optimizing their outreach, helping to allocate resources more effectively to those most in need. This can align with Jewish principles of *tzedakah* (charity) and justice, ensuring that resources are distributed in a way that promotes human dignity and addresses the systemic barriers that prevent equal access to basic needs.

AI can also be a tool for fostering dialogue and understanding between diverse religious communities. Interfaith platforms powered by AI could facilitate meaningful conversations between people of different faiths, promoting peace and mutual respect in an increasingly polarized world. These platforms could analyze the teachings of various religious traditions, identifying common values and helping to bridge divides. In this way, AI could contribute to the *shalom bayit*—the peace of the household—on a global scale.

A Note of Caution: Ethical Dilemmas and Spiritual Risks

While the potential benefits of AI are vast, we must approach its development and integration into society with great caution. The very technologies that offer hope can also bring unintended consequences if left unchecked. Jewish tradition teaches us the value of discernment. The Book of Proverbs reminds us that "the beginning of wisdom is the fear of the Lord" (Proverbs 9:10), suggesting that wisdom requires humility and reverence for the moral and ethical boundaries set by God. It is in this spirit of discernment that we must address the potential dangers AI poses to faith and society.

1. The Challenge of Autonomy and Human Dignity

A major ethical concern surrounding AI is its potential to undermine human autonomy and dignity. Judaism holds that each human being is created *b'tzelem Elohim* (in the image of God) and endowed with free will. This belief underscores the inherent dignity of every individual and the responsibility each person has to make moral choices.

As AI systems become more advanced, particularly in decision-making capacities, there is a risk that human agency could be diminished. Autonomous AI systems might make decisions in areas like healthcare, criminal justice, or employment, raising questions about accountability and the preservation of human dignity. If AI is allowed to make life-altering decisions without human oversight, we risk creating a society in which individuals are no longer seen as agents of their destiny but as subjects of impersonal algorithms.

This caution is especially relevant in religious contexts. Judaism teaches that the relationship between God and humanity is one of covenant, wherein individuals are called to make moral decisions that reflect their commitment to divine commandments. If AI systems begin to take over aspects of decision-making, particularly in moral or ethical realms, it could interfere with this covenantal relationship. As a community, we must ensure that AI enhances rather than erodes human moral agency.

As Rabbi Joseph B. Soloveitchik writes in *The Lonely Man of Faith*, "Man stands before God as a free agent. He is called upon to make his own decisions and to act in accordance with divine guidance. Any technological development that undermines this autonomy must be approached with extreme caution."[4] Rabbi Soloveitchik's words serve as a reminder that human free will is central to the religious experience, and that we must guard against any technology that diminishes this essential aspect of our humanity.

2. The Risk of Dehumanization

There is also the danger that AI, in its quest for efficiency and optimization, could lead to the dehumanization of society. When we begin to see people as data points or inputs in an algorithm, we lose sight of the divine image in each person. This is particularly concerning in contexts like employment, where AI-driven hiring algorithms could reduce individuals to mere qualifications or patterns, ignoring their intrinsic worth as human beings.

In religious communities, where the value of each soul is paramount, this risk is especially troubling. How do we ensure that AI respects the dignity and worth of every person, especially those who might be marginalized by technological systems? Jewish ethics demands that we place human dignity at the center of all technological advancements, ensuring that AI is always a servant to human flourishing rather than a force of dehumanization.

This caution echoes the warning of Neil Postman, a secular scholar who often critiqued the impact of technology on human culture. In "Technopoly", Postman warns of the dangers of allowing technology to control every aspect of life: "Technological change is not additive; it is ecological. New technology does not merely add something; it changes everything."[5] AI has the potential to reshape human life in ways that could harm our sense of community and personhood, and this danger must be mitigated by ensuring that ethical considerations take precedence over technological advancement.

3. AI and Spiritual Life: The Risk of Distraction

Finally, there is the risk that AI, rather than deepening our spiritual lives, could serve as a distraction from them. The constant presence of technology in our lives has already led to what some have called the "distraction economy," where our attention is constantly pulled in multiple directions by digital devices. If AI becomes more integrated into our daily lives, there is a risk that it could exacerbate this trend, making it harder for individuals to find the quiet reflection and focus needed for spiritual growth.

Jewish tradition places a high value on *kavanah*—intention and focus in prayer and religious practice. AI-driven technologies, particularly those designed to maximize engagement, might make it harder for individuals to cultivate the inner stillness necessary for spiritual development. Rabbi Abraham Joshua Heschel, in his work *The Sabbath*, emphasizes the importance of sacred time and the need to step away from the distractions of the material world to connect with the divine. "We must not only be careful of how we use our time," he writes, "but also how we allow time to use us."[6] AI, if not carefully managed, could erode the sacred time needed for reflection, prayer, and deep spiritual engagement.

Toward a Harmonious Future: The Role of Faith in Shaping AI's Development

As we envision a future where AI and faith coexist harmoniously, it is crucial that religious communities take an active role in shaping the development and deployment of AI technologies. Jewish tradition teaches that wisdom comes from a combination of *chochmah* (knowledge) and *binah* (understanding), and it is with this combination that we must approach the ethical dilemmas posed by AI. Religious leaders and ethicists have a responsibility to engage with AI developers, ensuring that the technologies being created align with the values of justice, compassion, and human dignity.

This engagement can take many forms. Faith communities might advocate for AI regulations that prioritize transparency and accountability, ensuring that AI systems are used in ways that promote social good and avoid harm. They could also serve as spaces for critical reflection on the moral and spiritual implications of AI, offering guidance to individuals and societies on how to navigate the challenges of a rapidly changing technological landscape.

In conclusion, while AI presents both tremendous hope and significant challenges, the future of AI and faith can be one of harmony if approached with caution, wisdom, and ethical foresight. It is my prayer that we, as a global community, will harness AI's potential for good while remaining vigilant in safeguarding the moral and spiritual dimensions of our lives. As the prophet Micah reminds us: "What does the Lord require of you? To act justly and to love mercy and to walk humbly with your God" (Micah 6:8). May we walk this path with care as we journey into the future together.

Footnotes

1. Jonathan Sacks, *Morality: Restoring the Common Good in Divided Times* (Basic Books, 2020), 75.
2. Jonathan Sacks, *Future Tense: Jews, Judaism, and Israel in the Twenty-First Century* (Schocken, 2009), 123.
3. Jonathan Sacks, *To Heal a Fractured World: The Ethics of Responsibility* (Schocken, 2005), 64.
4. Joseph B. Soloveitchik, *The Lonely Man of Faith* (Doubleday, 2006), 110.
5. Neil Postman, *Technopoly: The Surrender of Culture to Technology* (Vintage, 1993), 18.
6. Abraham Joshua Heschel, *The Sabbath: Its Meaning for Modern Man* (Farrar, Straus and Giroux, 1951), 24.

Chapter 12: AI, Privacy, and Surveillance: A Jewish Ethical Response

The rapid expansion of AI-driven surveillance technologies has raised significant ethical concerns in the modern world. As artificial intelligence systems collect vast amounts of data—through facial recognition, location tracking, and data mining—issues surrounding privacy, autonomy, and dignity come to the forefront. In this chapter, we will explore these growing concerns through a Jewish ethical lens, drawing on centuries-old principles that remain relevant in our technologically advanced society.

Judaism has long valued the sanctity of privacy, the dignity of the individual, and the importance of ethical behavior, particularly as it relates to speech and information. Through the teachings of *Shmirat HaLashon* (guarding one's speech) and *Lashon Hara* (harmful speech), we will examine how Jewish wisdom addresses concerns around privacy, the misuse of personal information, and surveillance. By understanding these principles, we can better navigate the moral and ethical dilemmas posed by AI-driven surveillance, ensuring that privacy and human dignity are preserved even in an era of unprecedented technological advancement.

The Scope of AI-Driven Surveillance: A Modern Dilemma

The proliferation of AI technologies has led to significant advances in surveillance capabilities. From facial recognition systems that can track individuals across cities, to data mining algorithms that collect, analyze, and predict human behavior, AI has drastically expanded the scope of surveillance. Governments, corporations, and private entities now have access to a wealth of information, allowing for enhanced security measures but also raising concerns about privacy violations.

In Jewish thought, the right to privacy is closely linked to the concept of human dignity. Each person is created *b'tzelem Elohim*—in the image of God (Genesis 1:27)—and thus possesses an inherent worth that must be respected. This concept emphasizes that individuals are entitled to a degree of privacy and autonomy, free from unjustified intrusion. When AI technologies are used to monitor and collect data on individuals without their consent, it risks violating this divine image, reducing people to mere data points in a surveillance system.

The key ethical dilemma here is balancing the benefits of AI-driven surveillance—such as increased security and crime prevention—with the need to protect individual privacy and dignity. While Jewish law acknowledges the importance of public safety, it also stresses the need to maintain respect for personal boundaries. As Rabbi Joseph B. Soloveitchik notes in *The Lonely Man of Faith*, the essence of human dignity lies in the individual's ability to choose and maintain control over their own life. If this autonomy is undermined by invasive surveillance technologies, we risk stripping people of their dignity and agency.[1]

Jewish law also offers insights into how private and public spaces are distinguished. The Talmudic concept of *hezek re'iyah*—the damage caused by seeing someone in their private space—illustrates how Jewish tradition has long recognized the importance of privacy. The Mishnah (Bava Batra 2a) discusses the requirement to construct walls between properties to prevent neighbors from intruding on each other's privacy, offering a legal basis for protecting individuals from unwanted observation.[2] This ancient principle can be applied to modern technologies like AI, which have the capacity to surveil individuals even in spaces that were once considered private.

Shmirat HaLashon and Lashon Hara: Ethical Speech and Information Sharing

The Jewish ethical framework for privacy and information-sharing is rooted in the teachings of *Shmirat HaLashon* and *Lashon Hara*. *Shmirat HaLashon*, meaning "guarding the tongue," emphasizes the need for individuals to be mindful of their speech, particularly when it comes to sharing sensitive or potentially harmful information. *Lashon Hara*, which refers to speaking negatively about others, is prohibited even when the information is true, because it can cause harm or damage a person's reputation.

While these principles originally pertained to verbal speech, they offer profound insights into the ethical challenges posed by AI-driven surveillance and data collection. In many ways, data collection is the digital equivalent of speech. When companies and governments gather personal data, they are, in effect, collecting and disseminating information about individuals. Just as Judaism forbids the careless or harmful sharing of information through *Lashon Hara*, it can be argued that the indiscriminate collection and sharing of data without proper consent or safeguards is an ethical violation. In both cases, the dignity and privacy of the individual must be protected.

The Chofetz Chaim, in his seminal work *Shmirat HaLashon*, emphasizes that the harm caused by *Lashon Hara* can be irreparable, especially when it comes to damaging someone's reputation or well-being.[3] This concept can be extended to AI surveillance: the potential harm caused by the unauthorized or unethical use of personal data can be equally damaging, if not more so, in the digital age. Once data is collected, it can be used to manipulate, control, or harm individuals in ways they may not even be aware of. This underscores the need for robust ethical frameworks to govern AI surveillance and data use.

Additionally, *Shmirat HaLashon* teaches us the importance of restraint and intentionality in speech. It warns against speaking impulsively or sharing information that may cause harm. This principle is particularly relevant in the context of AI technologies, where data collection often happens without individuals being fully aware of how their information is being used. The vast quantities of data gathered by AI systems can easily be shared or leaked, resulting in unintended harm to individuals, just as *Lashon Hara* can cause harm when spoken thoughtlessly.

Tzelem Elohim and Human Dignity: The Case for Privacy

At the core of Jewish ethics is the belief that every person is created *b'tzelem Elohim*, or in the image of God. This belief confers upon each individual an inherent dignity that must be respected and protected. The concept of *Tzelem Elohim* emphasizes the sanctity of the individual and the moral imperative to treat others with respect, compassion, and fairness.

In the context of AI and surveillance, the principle of *Tzelem Elohim* raises important questions about the extent to which technology can intrude upon personal privacy. If we believe that each person carries a divine spark, then we must also believe that individuals are entitled to a sphere of personal autonomy and privacy. The Talmud underscores this idea by discussing the importance of personal space and the right to be free from unwanted intrusion, such as in *Bava Batra* 2a, which discusses the construction of privacy walls between neighbors.[4]

In the digital age, where AI technologies can track, monitor, and collect data on individuals without their knowledge, the concept of *Tzelem Elohim* demands that we consider the ethical implications of such surveillance. Privacy is not merely a legal right, but a moral imperative rooted in human dignity. When surveillance systems invade this private space, they undermine the divine image within each person.

The Torah offers additional guidance on the importance of privacy. For example, the mitzvah to place a parapet on one's roof (Deuteronomy 22:8) is not just a safety measure; it is a commandment that highlights the responsibility we must protect others from harm. In the context of AI surveillance, this can be understood as a call to build ethical safeguards that protect individuals from the harmful effects of invasive data collection. Jewish law frequently emphasizes *lifnim mishurat ha-din*—going beyond the letter of the law to ensure ethical conduct. Even when data collection is legal, it must be scrutinized to ensure it aligns with ethical principles.[5]

Balancing Security and Privacy: A Jewish Perspective

Jewish law recognizes the importance of public safety and security, but it also demands a careful balance between security measures and the protection of individual rights. The principle of *pikuach nefesh*—the saving of a life—permits the violation of certain commandments in order to protect human life (Mishnah Yoma 8:6). This principle could be invoked to justify the use of AI-driven surveillance in cases where it is necessary to prevent violence, terrorism, or crime. However, *pikuach nefesh* does not grant unlimited authority to infringe upon individual privacy.

The Mishnah teaches that even when public safety is at stake, measures must be taken to limit harm and avoid unnecessary violations of individual rights. In the case of AI surveillance, this suggests that while security is important, it must be implemented in a way that respects human dignity and minimizes the potential for harm or abuse. Surveillance systems should be transparent, accountable, and subject to oversight to ensure that they are not used for unethical purposes, such as discrimination, oppression, or unjust control over populations.

The Rambam (Maimonides), in his *Mishnah Torah*, emphasizes that any violation of privacy or property rights must be justified by a compelling reason, such as the prevention of harm or the protection of public safety.[6] This teaching can guide our approach to AI surveillance: while there may be legitimate reasons to use surveillance technologies, they must be carefully regulated to prevent unnecessary harm or violations of individual rights.

Rabbi Lord Jonathan Sacks, in his discussion of ethics and technology, wrote that while technology can enhance human life, it must be guided by moral values to ensure that it does not infringe on fundamental human rights. "Technology gives us power," he writes, "but without ethics, it can easily turn into oppression."[7] This warning is particularly relevant to AI surveillance, which, if left unchecked, can easily become a tool for authoritarian control or the erosion of civil liberties.

The Responsibility of Companies, Governments, and Individuals

Considering these ethical concerns, it is incumbent upon companies, governments, and individuals to create and enforce ethical frameworks for AI surveillance. Jewish law emphasizes the concept of *arevut*—mutual responsibility—where each member of society is responsible for the welfare of others. This principle can be applied to the development and use of AI technologies, with stakeholders across various sectors holding a shared responsibility to ensure that AI is used in a way that respects human dignity and privacy.

Corporations that develop and deploy AI surveillance systems must prioritize transparency and accountability. This means providing clear information about how data is collected, stored, and used, as well as offering individuals the ability to control their own data. Ethical guidelines should also be in place to prevent misuse, such as discrimination or profiling, based on collected data.

Governments, too, have a responsibility to enact legislation that protects individuals from unwarranted surveillance. Just as Jewish law includes numerous safeguards to protect individuals from harm, modern legal systems must include protections against the misuse of AI technologies. This includes the right to privacy, the right to be informed about data collection, and the right to challenge unlawful surveillance.

Finally, individuals themselves have a role to play. Just as *Shmirat HaLashon* teaches us to guard our speech and be mindful of the information we share; individuals must be vigilant about their own digital privacy. This means being aware of the data we provide online, advocating for ethical AI practices, and holding companies and governments accountable when privacy violations occur.

Conclusion: Building an Ethical Future

As AI technologies continue to evolve, the ethical questions surrounding privacy and surveillance will only become more pressing. Jewish ethics, with its emphasis on human dignity, privacy, and the sanctity of the individual, provides valuable guidance for navigating these challenges. By drawing on the principles of *Shmirat HaLashon*, *Lashon Hara*, and *Tzelem Elohim*, we can develop ethical frameworks that protect privacy while also allowing for the benefits of AI-driven technologies.

The balance between security and privacy is not an easy one, but Jewish teachings remind us that human dignity must always be central to our decisions. As we move forward in this digital age, it is essential that we continue to uphold these values, ensuring that AI serves humanity without undermining the ethical foundations that protect human freedom and dignity.

Footnotes

1. Joseph B. Soloveitchik, *The Lonely Man of Faith* (Doubleday, 2006), 110.
2. *Mishnah Bava Batra* 2a, Talmud Bavli.
3. Israel Meir Kagan (Chofetz Chaim), *Shmirat HaLashon*, Part 1, Chapter 4.
4. *Mishnah Bava Batra* 2a, Talmud Bavli.
5. Deuteronomy 22:8; See Rambam's *Mishnah Torah*, Hilchot Nizkei Mamon, Chapter 5.
6. Rambam, *Mishnah Torah*, Hilchot Nizkei Mamon, Chapter 6.
7. Jonathan Sacks, *Future Tense: Jews, Judaism, and Israel in the Twenty-First Century* (Schocken, 2009), 154.

Chapter 13: AI and Education: Transforming Torah Study and Religious Learning

Artificial intelligence (AI) is reshaping various fields, and education is no exception. Within the realm of Jewish education, the traditional study of Torah, Talmud, and other sacred texts is undergoing a transformation through the integration of AI technologies. The ability to analyze vast amounts of textual data, offer personalized learning pathways, and provide instant access to commentaries has opened new avenues for Torah study. In this chapter, we explore how AI tools are revolutionizing the study of sacred texts while remaining mindful of the risks associated with over-reliance on technology. As a conservative rabbi, I believe it is essential to examine AI's role in religious learning through the lens of Jewish tradition, drawing on concepts like *chevruta* (partnership in study) and *masorah* (the transmission of tradition) that have been the traditional methods of study since the Mishna (200 CE), Talmud (500 CE) and the academies of Sura and Pumbedita (approximately 233-1033 CE), that were located in Babylonia, which is located in modern-day Iraq.

The Rise of AI in Jewish Education

Jewish education has always placed a premium on the study of sacred texts, with Torah study forming the bedrock of Jewish intellectual and spiritual life. This is the reason that Jews have always been known as "the people of the book." Traditionally, learning has taken place in yeshivot, synagogues, and schools where students engage deeply with the texts under the guidance of melamdim (teachers) and rabbis. In recent years, however, AI has begun to play an increasingly prominent role in Jewish education, particularly in the way we approach Torah study.

1. AI Tools for Torah Study

One of the most significant advancements in AI-driven education is the development of digital platforms that make Torah study more accessible. Applications like Sefaria, Bar Ilan's Responsa Project, and others use AI to provide comprehensive databases of Jewish texts, searchable commentaries, and interactive study tools. These platforms allow students to access the full range of Jewish texts in one place, from the Tanakh to Talmud, Midrash, and the writings of the great medieval and modern commentators.[1]

AI tools also enable automated commentary analysis, allowing students to quickly compare the interpretations of Rashi, Ramban, Ibn Ezra, and others with a few clicks. Machine learning algorithms can even identify thematic links between texts and suggest interpretations that align with or differ from traditional views. For instance, a student studying the laws of Shabbat in the Mishnah might be prompted with related discussions in the Talmud, responsa literature, or contemporary halachic rulings, enhancing their understanding of how Jewish law has evolved over time.

This technological advancement is a major shift from the traditional methods of Torah study that required students to have access to physical copies of texts or, in many cases, memorize large portions. The democratization of Jewish learning through AI tools has reduced barriers to entry, making Jewish scholarship accessible to a broader audience, including those outside traditional learning institutions.[2] While this is a tremendous achievement, it raises questions about how these changes may impact the way Torah study is approached and understood. Furthermore, will it become about the monetization of knowledge or knowledge for the sake of education? AI might erase the line between the two. Thus, lessening Torah, l'shem shamayim, or learning for the sake of heaven.

2. Personalized Learning Pathways

Another way AI is transforming Torah study is through personalized learning pathways. Much like in secular education, AI can assess a student's progress and adjust the learning material to fit their unique needs. If a student is struggling with a particular section of the Gemara, AI-driven platforms can provide additional resources, alternative explanations, or step-by-step breakdowns to help them grasp the material. Conversely, if a student excels, AI can suggest more advanced topics or related texts to explore, ensuring that learning remains challenging and engaging.

This personalized approach mirrors the individualized attention that a student might receive in a traditional *chevruta* or teacher-student relationship, where the more experienced partner adjusts the learning process to help the less experienced partner succeed.[3] However, it raises important questions about the role of human educators in an increasingly AI-driven learning environment. While AI can provide information, it lacks the ability to impart the ethical, spiritual, and emotional wisdom that is transmitted through human interaction.

The Value of *Chevruta* and Human Mentorship

One of the central concerns regarding AI in religious education is the risk of over-reliance on technology at the expense of human interaction and mentorship. Jewish tradition places great value on *chevruta*—the partnership between two people in study. I know that my rabbinate has been greatly enhanced with my daily study with my teacher, Rabbi Sidney Zimelman. In the classical model of Yeshiva study, students work together, debating the meaning of texts, raising questions, and challenging one another's interpretations. This dialectical process not only deepens the students' understanding of the material but also fosters personal growth, humility, and mutual respect.

Within this context, Rabbi Joseph B. Soloveitchik emphasizes the importance of *chevruta* as a means of cultivating intellectual and spiritual virtues. In his view, the back-and-forth exchange of ideas between study partners is a crucial element of Torah study because it replicates the dynamic interaction between humanity and God in the pursuit of truth.[4] AI, while capable of providing vast amounts of information, lacks the relational depth and spontaneity that *chevruta* offers.

Additionally, the role of teachers and rabbis as guides and mentors cannot be overstated. The transmission of Jewish knowledge is not merely about acquiring information; it is about shaping the moral and ethical character of students. This is the essence of *masorah*—the transmission of tradition from generation to generation. As the Mishnah in *Pirkei Avot* teaches, "Moses received the Torah at Sinai and handed it down to Joshua, Joshua to the elders, the elders to the prophets, and the prophets handed it down to the men of the Great Assembly" (Pirkei Avot 1:1). This chain of tradition highlights the interpersonal nature of Jewish learning, where wisdom is passed from teacher to student, infused with the values and lived experiences of the teacher.[5]

The risk of AI-driven learning environments is that they may diminish the role of human educators, turning Torah study into a solitary activity focused solely on the acquisition of knowledge. While AI can enhance learning, it should never replace the mentorship and guidance that are essential to Jewish education. The relationship between student and teacher, and the dialogical nature of Torah study, are indispensable for developing not only knowledge but also character, ethics, and spiritual insight.

The Potential Pitfalls of AI in Religious Education

While AI holds great promise for enhancing Torah study, it also presents certain risks and potential drawbacks. One major concern is that AI, with its emphasis on efficiency and personalization, may inadvertently encourage a superficial approach to learning. The depth of Torah study lies not in how quickly one can access information but in the process of grappling with the text, encountering ambiguity, and wrestling with difficult questions. Rabbi Abraham Joshua Heschel cautions against an overly mechanistic approach to education, warning that "the danger is not that we will lose our way in the jungle of knowledge, but that we may fail to hear the voice of wonder."[6] In the context of AI, there is a risk that students may prioritize speed and convenience over the deep, reflective study that is central to Jewish learning. Thus, losing out on the deeper meanings of the text.

Another potential drawback is the temptation to rely too heavily on AI-generated insights and commentaries, at the expense of developing one's own interpretative skills. Traditional Torah study requires active engagement, critical thinking, and creativity. Students are encouraged to ask questions, challenge assumptions, and develop their own understanding of the text. If students come to depend on AI to provide ready-made answers, they may lose the ability to think independently and cultivate the skills necessary for meaningful Torah study.[7]

Furthermore, there are ethical concerns regarding the data-driven nature of AI. Many AI tools rely on the collection and analysis of data to personalize learning experiences. In Jewish tradition, privacy and the dignity of the individual are paramount, as reflected in the principle of *Tzelem Elohim* (the belief that every human being is created in the image of God). While AI's data-driven approach can be beneficial for personalized learning, we must ensure that these technologies are used ethically, and that students' personal information is protected. As discussed in previous chapters, the balance between technological innovation and ethical responsibility is crucial in any application of AI.[8]

Integrating AI with Tradition: A Balanced Approach

Given the profound potential of AI in religious education, it is essential to adopt a balanced approach that integrates technology with the timeless values of Jewish tradition. Rather than viewing AI as a replacement for traditional forms of Torah study, we should see it as a tool that can complement and enhance the learning process.

One way to achieve this balance is by incorporating AI into *chevruta* study. For example, students could use AI tools to access commentaries and explore various interpretations during their study sessions, but the heart of the learning would remain the dialogical interaction between study partners. In this way, AI can provide valuable resources while preserving the relational and interactive aspects of *chevruta*.[9]

Similarly, teachers and rabbis can use AI to supplement their instruction, providing students with additional resources, interactive learning experiences, and personalized feedback. However, the role of the teacher as a mentor and guide must remain central. As *Pirkei Avot* teaches, "Make for yourself a teacher and acquire for yourself a friend" (Pirkei Avot 1:6).[10] This teaching underscores the importance of human relationships in learning. AI can assist in the process, but it cannot replace the wisdom, empathy, and moral guidance that a teacher provides.

In terms of *masorah*, AI can play a role in preserving and transmitting Jewish knowledge by digitizing and archiving sacred texts, ensuring that future generations have access to the full corpus of Jewish learning. However, *masorah* is not simply about the preservation of information; it is about the transmission of values, ethics, and a sense of responsibility to the past and the future. Human educators play a vital role in ensuring that students not only learn the texts but also understand their significance in the broader context of Jewish life.[11]

Conclusion: The Future of AI in Torah Study

As AI continues to develop and play a larger role in education, it is imperative that we approach its integration into Torah study with both optimism and caution. AI has the potential to revolutionize Jewish education, making Torah study more accessible, interactive, and personalized. It can offer new ways of engaging with sacred texts, helping students to deepen their knowledge and enhance their understanding.

However, we must also remain mindful of the limitations of AI. Torah study is not merely an intellectual exercise; it is a spiritual and moral pursuit that requires human interaction, mentorship, and reflection. The relational and communal aspects of Jewish learning, embodied in the concepts of *chevruta* and *masorah*, are essential to the transmission of Jewish wisdom. As we embrace the possibilities of AI, we must ensure that it remains a tool that enhances, rather than replaces, the irreplaceable human elements of Jewish education.

Footnotes

1. Sefaria: A Living Library of Jewish Texts Online. www.sefaria.org.
2. Ruth Calderon, *A Bride for One Night: Talmud Tales* (Jewish Publication Society, 2014).
3. Jonathan Rosen, *The Talmud and the Internet: A Journey between Worlds* (Farrar, Straus, and Giroux, 2000), 95.
4. Joseph B. Soloveitchik, *The Lonely Man of Faith* (Doubleday, 2006), 109. ↩
5. *Pirkei Avot* 1:1, *Mishnah*.
6. Abraham Joshua Heschel, *Man Is Not Alone: A Philosophy of Religion* (Farrar, Straus, and Giroux, 1951), 89.
7. Avi Warshavsky, "Can Artificial Intelligence Replace the Rabbi? A Halachic Perspective on AI and Torah Learning," *Tradition: A Journal of Orthodox Jewish Thought*, Spring 2023.
8. Shoshana Zuboff, *The Age of Surveillance Capitalism: The Fight for a Human Future at the New Frontier of Power* (PublicAffairs, 2019).
9. Erica Brown, *Spiritual Boredom: Rediscovering the Wonder of Judaism* (Jewish Lights Publishing, 2009), 125.
10. *Pirkei Avot* 1:6, *Mishnah*.
11. Ismar Schorsch, *Canon Without Closure: Torah Commentary* (Jewish Theological Seminary Press, 2007).

Chapter 14: AI and Interfaith Dialogue: Building Bridges Between Traditions

As artificial intelligence (AI) increasingly influences various sectors of society, its potential to foster interfaith dialogue has become a subject of significant interest. The use of AI to facilitate conversations and promote understanding between different religious traditions offers an unprecedented opportunity to build bridges across faith communities. Such a possibility is tremendously important in today's divided political, religious and cultural landscape. This chapter will explore how AI can serve as a powerful tool in this endeavor, providing platforms for shared learning, analyzing religious texts, and uncovering common ethical themes that can serve as a foundation for mutual respect and collaboration.

From a Jewish perspective, the theological imperatives of *Kiddush Hashem* (sanctification of God's name) and *Darchei Shalom* (ways of peace) offer valuable frameworks for understanding the role of AI in promoting peace and mutual understanding. These concepts, deeply embedded in Jewish tradition, emphasize the importance of fostering harmony, justice, and ethical behavior within and between communities. By drawing on these teachings, we can better appreciate the ways AI can contribute to the larger mission of *Tikkun Olam*—repairing the world—through interfaith dialogue and collaboration.

AI's Role in Facilitating Interfaith Dialogue

The emergence of AI-powered platforms designed to support interfaith dialogue marks a significant shift in how religious communities can engage with one another. Traditionally, interfaith discussions have been limited by geography, language barriers, and cultural differences. However, AI can transcend these limitations by providing digital spaces where religious leaders, scholars, and laypeople can engage in meaningful conversations across vast distances. AI tools can facilitate real-time translations, analyze vast religious texts, and identify common ethical principles across different faith traditions.

1. Shared Learning Platforms

AI-driven platforms are emerging as vital tools for promoting shared learning across religious boundaries. These platforms allow for the comparative study of sacred texts from multiple faiths, enabling religious scholars and students to explore common themes and differences in religious teachings. For example, AI algorithms can analyze the Torah, New Testament, the Qur'an, and other religious texts, identifying shared ethical concerns such as the promotion of justice, charity, and compassion.

This capability aligns with the Jewish concept of *Kiddush Hashem*, which calls for actions that sanctify God's name through righteous behavior in the presence of others. Engaging in respectful interfaith dialogue and exploring shared ethical concerns are ways in which religious communities can come together to promote peace and understanding. AI's ability to facilitate these interactions enhances the potential for such dialogue to yield fruitful and meaningful outcomes.

For example, a platform might bring together a Jewish scholar studying *Tzedek* (justice), a Christian theologian exploring social justice in the Gospels, and a Muslim leader discussing *Zakat* (charity) in Islamic teachings. The platform could analyze these discussions in real-time, offering insights into how different traditions conceptualize justice and suggesting areas for further exploration. This creates an environment where participants can engage with one another respectfully and constructively, reinforcing the idea that while doctrinal differences exist, the ethical foundations of many religions often hold commonalities.

2. Automated Textual Analysis Across Traditions

AI's capacity for analyzing vast amounts of textual data has significant implications for interfaith dialogue. Machine learning algorithms can sift through religious texts, commentary, and historical writings from multiple traditions to identify common ethical themes, philosophical ideas, and shared narratives. This kind of analysis can help religious communities recognize the common ground they share, even when their theological frameworks differ. This being incredibly important in the time of religious strife and tribalization of religion.

For instance, AI could be used to conduct a comparative analysis of Jewish, Christian, and Islamic teachings on charity. In Judaism, *tzedakah* is not merely a voluntary act of giving but a legal obligation that embodies the principle of justice. Similarly, in Christianity, the concept of charity is deeply rooted in the teachings of Jesus, particularly in his Sermon on the Mount. In Islam, the principle of *zakat* mandates charitable giving as one of the Five Pillars of the faith. AI can illuminate these parallels, offering a rich foundation for interfaith discussions on how religious communities can work together to address global issues like poverty and economic inequality.

The Talmud teaches that *Darchei Shalom*—the ways of peace—are essential for maintaining harmony between individuals and communities. The idea is that peace is not merely the absence of conflict but the active pursuit of justice and understanding. By highlighting shared ethical concerns through AI-driven textual analysis, religious communities can take concrete steps toward building a more just and harmonious world.

Jewish Teachings on *Kiddush Hashem* and *Darchei Shalom*

Jewish tradition provides a strong theological basis for using technology, including AI, to promote peace and mutual understanding between religious communities. Two key concepts—*Kiddush Hashem* and *Darchei Shalom*—offer insights into how AI can be harnessed ethically to foster interfaith dialogue.

1. *Kiddush Hashem*: Sanctifying God's Name

Kiddush Hashem refers to actions that sanctify God's name by reflecting Jewish values of justice, compassion, and ethical behavior in the world. Traditionally, this concept has been applied to acts of moral courage, ethical integrity, and kindness that inspire others and reflect positively on Am Yisrael (the Jewish people), Judaism (the religion) and individual Jews and our relationship with God.

In the context of AI and interfaith dialogue, *Kiddush Hashem* can be seen in efforts to use technology for the greater good, promoting understanding and collaboration between different faith communities. When Jewish leaders and scholars engage in interfaith dialogue through AI platforms, they participate in sanctifying God's name by demonstrating that religious traditions can work together to promote justice, peace, and human dignity.

Rabbi Jonathan Sacks, of blessed memory, emphasized the importance of using modern tools to further ethical goals. In his work *The Dignity of Difference*, Rabbi Sacks argued that we should not only tolerate religious diversity but actively celebrate it, recognizing that different faiths can offer valuable perspectives on the universal human pursuit of meaning and morality.[1] By employing AI to engage in interfaith dialogue, religious communities uphold the value of *Kiddush Hashem* by demonstrating how technology can be used to bring people together rather than divide them.

2. *Darchei Shalom*: The Ways of Peace

Darchei Shalom, or "the ways of peace," is a Talmudic principle that promotes harmonious relationships between Jews and non-Jews. The Talmud outlines numerous laws designed to foster peace between neighbors, regardless of their faith, recognizing the importance of maintaining good relations in a diverse society.

In the modern context, *Darchei Shalom* extends to the global community, encouraging Jews to engage constructively with people of other faiths and cultures. AI, when used to promote interfaith dialogue and collaboration, aligns with this principle by offering a platform for peaceful and respectful interactions between religious traditions. By fostering understanding and promoting shared values, AI-driven initiatives can help reduce tension and conflict between religious communities.

The Rambam (Maimonides) reinforced the importance of peaceful coexistence when he wrote, "The Torah was given only to bring peace to the world" (Mishnah Torah, Hilchot Chanukah 4:14).[2] This teaching can be applied to AI's role in interfaith dialogue: just as the Torah seeks to bring peace, so too can AI be a tool for achieving this virtuous goal by facilitating communication and collaboration across faith boundaries.

Case Studies: AI in Interfaith Collaboration

Several recent initiatives demonstrate how AI has been used to foster interfaith dialogue and collaboration in addressing global challenges. These case studies highlight the potential of AI to promote peace and mutual understanding between religious traditions.

1. The Abrahamic Faiths AI Project

In 2022, an initiative known as the Abrahamic Faiths AI Project brought together Jewish, Christian, and Muslim scholars to explore how AI could be used to promote interfaith understanding. The project utilized AI-driven platforms to analyze religious texts from the Torah, Bible, and Qur'an, focusing on common ethical teachings related to social justice, charity, and environmental stewardship.

The project also used AI to facilitate online discussions between religious leaders, allowing them to share insights and collaborate on initiatives addressing issues such as climate change and poverty. This collaboration led to the creation of a joint statement from Jewish, Christian, and Muslim leaders calling for greater interfaith cooperation in addressing environmental challenges, demonstrating how AI can be a catalyst for global action based on shared values.

2. AI for Social Justice: An Interfaith Approach

Another example of AI-driven interfaith collaboration is the AI for Social Justice initiative, which brings together religious leaders from various faith traditions to address issues of economic inequality and human rights. The initiative uses AI to analyze data related to global poverty, identifying areas where interfaith partnerships can make a tangible difference.

Through this project, Jewish, Christian, Muslim, and Buddhist leaders have collaborated on initiatives to provide food, medical care, and educational resources to underserved communities. The success of this project demonstrates how AI can be used to not only facilitate dialogue but also drive concrete action based on shared religious values.

Ethical Guidelines for AI in Interfaith Dialogue

While the potential for AI to foster interfaith dialogue is significant, it is essential to approach this technology with ethical considerations in mind. Religious communities must ensure that AI is used responsibly, promoting peace and mutual understanding without reinforcing stereotypes or perpetuating conflict.

1. Transparency and Accountability

AI-driven platforms used for interfaith dialogue must be transparent in their algorithms and data usage. Religious leaders and communities should be informed about how AI systems work, ensuring that these technologies are used in a way that promotes trust and accountability. AI should be designed to encourage constructive and respectful dialogue, avoiding the amplification of divisive content or misinformation.

2. Respect for Religious Traditions

AI platforms must respect the diversity of religious traditions, recognizing the importance of doctrinal differences while emphasizing shared ethical principles. In interfaith dialogue, it is essential to approach each tradition with humility and a willingness to learn. AI can facilitate this process by highlighting commonalities, but it should never be used to diminish or oversimplify the richness of individual religious traditions.

3. Collaboration and Inclusivity

Successful interfaith dialogue requires collaboration and inclusivity. AI platforms should be designed to include voices from a wide range of religious traditions, ensuring that minority faiths are represented alongside more prominent traditions. By fostering a spirit of inclusivity, AI can help create a more equitable and just dialogue that reflects the diversity of the global religious community.

Conclusion: AI as a Tool for Peace and Unity

As AI continues to evolve, its potential to foster interfaith dialogue and collaboration offers a unique opportunity to build bridges between religious traditions. By providing platforms for shared learning, analyzing religious texts, and promoting understanding between communities, AI can help create a more peaceful and just world.

Jewish teachings on *Kiddush Hashem* and *Darchei Shalom* remind us that the pursuit of peace and justice is not only a religious obligation but also a moral imperative. As religious communities harness the power of AI to engage in interfaith dialogue, they can contribute to the larger mission of *Tikkun Olam*, working together to address global challenges and promote the well-being of all people.

Footnotes

1. Jonathan Sacks, *The Dignity of Difference: How to Avoid the Clash of Civilizations* (Continuum, 2002), 19.
2. Maimonides, *Mishnah Torah*, Hilchot Chanukah 4:14.

Chapter 15: Thoughts on Bridging Technology and Theology for a Better Tomorrow

The rapid development of artificial intelligence has brought profound changes to many aspects of modern life, including the way we approach religion, spirituality, and theology. I know that it has for me. When I began my rabbinate over a quarter of a century ago, AI wasn't even thought of within the religious world or rabbinic training, especially not in Israel, and not in mine. As this book has explored, AI has introduced new opportunities for deepening our understanding of sacred texts, fostering interfaith dialogue, and addressing global challenges through ethical and social justice initiatives. At the same time, it has raised critical questions about the limits of technology, the preservation of human agency, and the ethical boundaries that must be carefully guarded.

From my perspective as a conservative rabbi, the conclusion is clear: AI has the potential to be a valuable tool in religious life, but it must be approached with discernment and humility. While AI can enhance our ability to engage with religious texts, analyze complex commentaries, and bring people of different faiths together, it must never replace the wisdom of our tradition, the authority of sacred texts, or the role of human mentorship and moral judgment. AI can offer insights and facilitate learning, but it cannot replicate the depth, nuance, and lived experience that come from millennia of interpretation, transmission, and spiritual growth.

Jewish tradition, particularly the mystical teachings of the *Zohar*, offers a framework for understanding the proper role of AI in our ongoing spiritual journey. The *Zohar* describes life and learning as a process of unfolding spiritual growth—a continuous journey toward greater understanding of the Divine. In this context, AI is merely a tool that can aid us on this journey. It is a tool that can illuminate new pathways of thought, uncover connections between texts, and make religious learning more accessible to broader audiences. But it is not the end of the journey, nor the destination itself. The journey of faith is, at its core, a deeply human experience—rooted in tradition, personal reflection, and communal engagement.

To be a Jew, or a religious person of any faith, is to be part of an ongoing journey that spans generations. As Rabbi Jonathan Sacks often emphasized, being religious is not about reaching a destination of certainty or perfection, but about joining the *derech*, the path that our people have walked for thousands of years. It is a path of inquiry, doubt, faith, struggle, and growth. AI can assist us in this process, but it cannot replace the essential human elements of that journey. Technology can offer knowledge, but it cannot replace wisdom, which is gained through lived experience, ethical choices, and communal life.

Rabbi Sacks also spoke of the importance of memory—of knowing where we come from and to whom we are responsible. The Jewish tradition is a repository of collective memory, stretching from Sinai to today, encompassing the voices of prophets, rabbis, scholars, and ordinary people who have shaped the story of our people. This memory is not something that can be digitized or automated. It is lived, shared, and passed down through *masorah*, the chain of tradition that connects past generations to the future. L'Dor V'Dor, From Generation to Generation.

In the end, AI is a tool—an advanced one—but still just a tool in our ongoing relationship with God. It can assist us in our studies and open new windows of understanding, but it cannot replace the essential human task of living out our faith, guided by the teachings of the Torah, the wisdom of our ancestors, and the values that have sustained us for millennia. The ethical dilemmas, spiritual questions, and communal challenges that AI presents are not to be feared but embraced, as part of our responsibility to steward the world and our tradition with care, humility, and moral courage.

As the *Zohar* reminds us, the journey of the soul is a continuous process, one that draws us ever closer to God. As we integrate AI into our religious lives, we must remember that it is just one more tool in this journey—a tool that can enhance our understanding but that will never replace the sacred and enduring task of being part of a living, breathing community of faith. To be a Jew is to be part of this journey, a journey that demands not just knowledge but wisdom, not just technology but humanity, and not just individual insight but collective responsibility. In this, we are reminded that the future of faith is in our hands, and that AI, like all tools, must be used to build, not replace, the ethical and spiritual framework that has sustained us throughout history. I know that it is a tool that I will use, but only one of many in the toolbox of rabbinic wisdom and future understandings of our codex of life and knowledge.

About the Author

Rabbi Andrew Bloom, a native of New Jersey, brings a diverse background to his rabbinic work. He served in the Israel Defense Forces (IDF) and holds a Bachelor of Education from Seminar Hakibutzim. Afterward, he received ordination and a master's degree from the Schechter Institute of Rabbinic Studies.

Now residing in Fort Worth, Texas, Rabbi Bloom has been deeply involved in community initiatives, including co-chairing the Fort Worth Task Force on Race and Culture and serving on various civic boards, such as the Mayor's Faith-Based Cabinet. As a co-founder of Cowtown Clergy, an author, and a certified ethicist with credentials from the University of Texas at Austin's McCombs School and the Wharton School, he is committed to bridging faith, ethics, and community engagement.

Rabbi Bloom is married to Michal, and together they have three children: Daniel, Maya, and Lia. He considers his family his greatest achievement.

www.ingramcontent.com/pod-product-compliance
Lightning Source LLC
Chambersburg PA
CBHW032051150426
43194CB00006B/493